A Comprehensive Handbook of Animal Science

A Comprehensive Handbook of Animal Science

Editor: Allan Bardsley

R CALLISTO REFERENCE

www.callistoreference.com

Callisto Reference,
118-35 Queens Blvd., Suite 400,
Forest Hills, NY 11375, USA

Visit us on the World Wide Web at:
www.callistoreference.com

ISBN: 978-1-64116-016-2 (Hardback)

Cataloging-in-Publication Data

A comprehensive handbook of animal science / edited by Allan Bardsley.
    p. cm.
Includes bibliographical references and index.
ISBN 978-1-64116-016-2
1. Zoology. 2. Animals. I. Bardsley, Allan.
QL45.2 .C66 2018
590--dc23

Table of Contents

Preface

Animal science refers to the study of animals, their DNA structure, organs, bodily functions, behavior and nutrition. It deals with all categories of animals like cattle, goats, horses, dogs, cats, sheep, pigs and other exotic species. The subject also includes specialized topics like dairy management, swine management, poultry production, small ruminant management, etc. This book presents the complex subject of animal science in the most comprehensible and easy to understand language. The topics included in it are of utmost significance and bound to provide incredible insights to readers. This textbook is an essential guide for both academicians and those who wish to pursue this discipline further.

Given below is the chapter wise description of the book:

Chapter 1- The subject which studies the biology of animals is known as animal science. The subject explores ways of treating animals and using them for the purpose of animal husbandry. This is an introductory chapter which will introduce briefly all the significant aspects of animal science.

Chapter 2- Animals can be classified into parazoa and eumetazoa. Parazoa consists of sponges and the Trichoplax adhaerens (Placozoa) whereas eumetazoa consists of radiate, bilateria, mesozoa and nephrozoa. This chapter has been carefully written to provide an easy understanding of the classification of animals.

Chapter 3- Vertebrates are animals which have a spinal cord whereas animals without a spine are referred to as invertebrates. Some of the vertebrate animals are fish and amphibian. The topics discussed in the chapter are of great importance to broaden the existing knowledge on vertebrate and invertebrate animals.

Chapter 4- Animal nutrition and diet studies the nutrition that animals require to stay healthy. Plants possess autotrophic mode of nutrition while animals need to consume food for nutrition. This chapter will provide an integrated understanding of animal science.

Indeed, my job was extremely crucial and challenging as I had to ensure that every chapter is informative and structured in a student-friendly manner. I am thankful for the support provided by my family and colleagues during the completion of this book.

Editor

An Introduction to Animal Science

The subject which studies the biology of animals is known as animal science. The subject explores ways of treating animals and using them for the purpose of animal husbandry. This is an introductory chapter which will introduce briefly all the significant aspects of animal science.

Animal

Animals are multicellular, eukaryotic organisms of the kingdom Animalia (also called Metazoa). The animal kingdom emerged as a clade within Apoikozoa as the sister group to the choanoflagellates. Animals are motile, meaning they can move spontaneously and independently at some point in their lives. Their body plan eventually becomes fixed as they develop, although some undergo a process of metamorphosis later in their lives. All animals are heterotrophs: they must ingest other organisms or their products for sustenance.

Most known animal phyla appeared in the fossil record as marine species during the Cambrian explosion, about 542 million years ago. Animals can be divided broadly into vertebrates and invertebrates. Vertebrates have a backbone or spine (vertebral column), and amount to less than five percent of all described animal species. They include fish, amphibians, reptiles, birds and mammals. The remaining animals are the invertebrates, which lack a backbone. These include molluscs (clams, oysters, octopuses, squid, snails); arthropods (millipedes, centipedes, insects, spiders, scorpions, crabs, lobsters, shrimp); annelids (earthworms, leeches), nematodes (filarial worms, hookworms), flatworms (tapeworms, liver flukes), cnidarians (jellyfish, sea anemones, corals), ctenophores (comb jellies), and sponges. The study of animals is called zoology.

Etymology

The word "animal" comes from the Latin *animalis*, meaning *having breath*, *having soul* or *living being*. In everyday non-scientific usage the word excludes humans – that is, animal is often used to refer only to non-human members of the kingdom Animalia; often, only closer relatives of humans such as mammals and other vertebrates, are meant. The biological definition of the word refers to all members of the kingdom Animalia, encompassing creatures as diverse as sponges, jellyfish, insects, and humans.

History of Classification

Aristotle divided the living world between animals and plants, and this was followed by Carl Linnaeus, in the first hierarchical classification. In Linnaeus's original scheme, the animals were one of three kingdoms, divided into the classes of Vermes, Insecta, Pisces, Amphibia, Aves, and Mammalia. Since then the last four have all been subsumed into a single phylum, the Chordata, whereas the various other forms have been separated out.

Carl Linnaeus, an animal himself, is known as the father of modern taxonomy.

In 1874, Ernst Haeckel divided the animal kingdom into two subkingdoms: Metazoa (multicellular animals) and Protozoa (single-celled animals). The protozoa were later moved to the kingdom Protista, leaving only the metazoa. Thus Metazoa is now considered a synonym of Animalia.

Characteristics

Animals have several characteristics that set them apart from other living things. Animals are eukaryotic and multicellular, which separates them from bacteria and most protists. They are heterotrophic, generally digesting food in an internal chamber, which separates them from plants and algae. They are also distinguished from plants, algae, and fungi by lacking rigid cell walls. All animals are motile, if only at certain life stages. In most animals, embryos pass through a blastula stage, which is a characteristic exclusive to animals.

Structure

With a few exceptions, most notably the sponges (Phylum Porifera) and Placozoa, animals have bodies differentiated into separate tissues. These include muscles, which are able to contract and control locomotion, and nerve tissues, which send and process signals. Typically, there is also an internal digestive chamber, with one or two openings. Animals with this sort of organization are called metazoans, or eumetazoans when the former is used for animals in general.

All animals have eukaryotic cells, surrounded by a characteristic extracellular matrix composed of collagen and elastic glycoproteins. This may be calcified to form structures like shells, bones, and spicules. During development, it forms a relatively flexible framework upon which cells can move about and be reorganized, making complex structures possible. In contrast, other multicellular organisms, like plants and fungi, have cells held in place by cell walls, and so develop by progressive growth. Also, unique to animal cells are the following intercellular junctions: tight junctions, gap junctions, and desmosomes.

Reproduction and Development

Nearly all animals undergo some form of sexual reproduction. They produce haploid gametes by meiosis. The smaller, motile gametes are spermatozoa and the larger, non-motile gametes are ova. These fuse to form zygotes, which develop into new individuals.

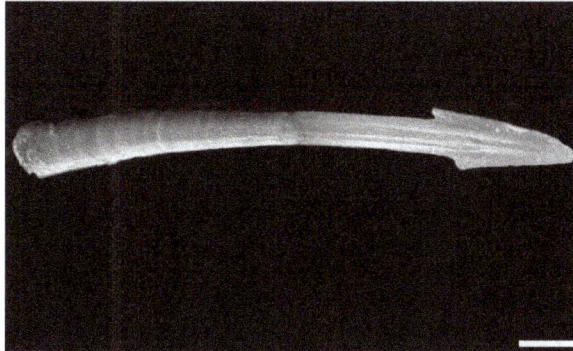

Some species of land snails use love darts as a form of sexual selection

Many animals are also capable of asexual reproduction. This may take place through parthenogenesis, where fertile eggs are produced without mating, budding, or fragmentation.

A zygote initially develops into a hollow sphere, called a blastula, which undergoes rearrangement and differentiation. In sponges, blastula larvae swim to a new location and develop into a new sponge. In most other groups, the blastula undergoes more complicated rearrangement. It first invaginates to form a gastrula with a digestive chamber, and two separate germ layers—an external ectoderm and an internal endoderm. In most cases, a mesoderm also develops between them. These germ layers then differentiate to form tissues and organs.

Inbreeding Avoidance

In Gombe Stream National Park, male chimpanzees remain in their natal community while females disperse to other groups

During sexual reproduction, mating with a close relative (inbreeding) generally leads to inbreeding depression. For instance, inbreeding was found to increase juvenile mortality in 11 small animal species. Inbreeding depression is considered to be largely due to expression of deleterious recessive mutations. Mating with unrelated or distantly related members of the same species is generally thought to provide the advantage of masking deleterious recessive mutations in progeny. Animals have evolved numerous diverse mechanisms for avoiding close inbreeding and promoting outcrossing.

As indicated in the image of chimpanzees, they have adopted dispersal as a way to separate close relatives and prevent inbreeding. Their dispersal route is known as natal dispersal, whereby individuals move away from the area of birth.

DNA analysis has shown that 60% of offspring in splendid fairywrens nests were sired through extra-pair copulations, rather than from resident males.

In various species, such as the splendid fairywren, females benefit by mating with multiple males, thus producing more offspring of higher genetic quality. Females that are pair bonded to a male of poor genetic quality, as is the case in inbreeding, are more likely to engage in extra-pair copulations in order to improve their reproductive success and the survivability of their offspring.

Food and Energy Sourcing

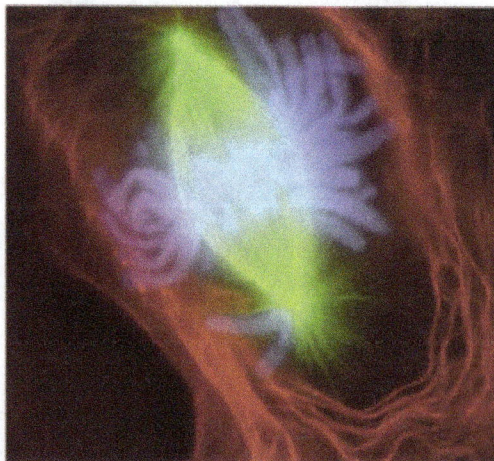

A newt lung cell stained with fluorescent dyes undergoing the early anaphase stage of mitosis

All animals are heterotrophs, meaning that they feed directly or indirectly on other living things. They are often further subdivided into groups such as carnivores, herbivores, omnivores, and parasites.

Predation is a biological interaction where a predator (a heterotroph that is hunting) feeds on its prey (the organism that is attacked). Predators may or may not kill their prey prior to feeding on them, but the act of predation almost always results in the death of the prey. The other main category of consumption is detritivory, the consumption of dead organic matter. It can at times be difficult to separate the two feeding behaviours, for example, where parasitic species prey on a host organism and then lay their eggs on it for their offspring to feed on its decaying corpse. Selective pressures imposed on one another has led to an evolutionary arms race between prey and predator, resulting in various antipredator adaptations.

Most animals indirectly use the energy of sunlight by eating plants or plant-eating animals. Most plants use light to convert inorganic molecules in their environment into carbohydrates, fats, proteins and other biomolecules, characteristically containing reduced carbon in the form of carbon-hydrogen bonds. Starting with carbon dioxide (CO_2) and water (H_2O), photosynthesis converts the energy of sunlight into chemical energy in the form of simple sugars (e.g., glucose), with the release of molecular oxygen. These sugars are then used as the building blocks for plant growth, including the production of other biomolecules. When an animal eats plants (or eats other animals which have eaten plants), the reduced carbon compounds in the food become a source of energy and building materials for the animal. They are either used directly to help the animal grow, or broken down, releasing stored solar energy, and giving the animal the energy required for motion.

Animals living close to hydrothermal vents and cold seeps on the ocean floor are not dependent on the energy of sunlight. Instead chemosynthetic archaea and bacteria form the base of the food chain.

Origin and Fossil Record

Dunkleosteus was a 10-metre-long (33 ft) prehistoric fish.

Animals are generally considered to have emerged within flagellated eukaryota. Their closest known living relatives are the choanoflagellates, collared flagellates that have a morphology similar to the choanocytes of certain sponges. Molecular studies place animals in a supergroup called the opisthokonts, which also include the choanoflagellates, fungi and a few small parasitic protists. The name comes from the posterior location of the flagellum in motile cells, such as most animal spermatozoa, whereas other eukaryotes tend to have anterior flagella.

The first fossils that might represent animals appear in the Trezona Formation at Trezona Bore, West Central Flinders, South Australia. These fossils are interpreted as being early sponges. They were found in 665-million-year-old rock.

The next oldest possible animal fossils are found towards the end of the Precambrian, around 610 million years ago, and are known as the Ediacaran or Vendian biota. These are difficult to relate to later fossils, however. Some may represent precursors of modern phyla, but they may be separate groups, and it is possible they are not really animals at all.

Aside from them, most known animal phyla make a more or less simultaneous appearance during the Cambrian period, about 542 million years ago. It is still disputed whether this event, called the Cambrian explosion, is due to a rapid divergence between different groups or due to a change in conditions that made fossilization possible.

Some palaeontologists suggest that animals appeared much earlier than the Cambrian explosion, possibly as early as 1 billion years ago. Trace fossils such as tracks and burrows found in the Tonian period indicate the presence of triploblastic worms, like metazoans, roughly as large (about 5 mm wide) and complex as earthworms. During the beginning of the Tonian period around 1 billion years ago, there was a decrease in Stromatolite diversity, which may indicate the appearance of grazing animals, since stromatolite diversity increased when grazing animals became extinct at the End Permian and End Ordovician extinction events, and decreased shortly after the grazer populations recovered. However the discovery that tracks very similar to these early trace fossils are produced today by the giant single-celled protist *Gromia sphaerica* casts doubt on their interpretation as evidence of early animal evolution.

Groups of Animals

Traditional morphological and modern molecular phylogenetic analysis have both recognized a major evolutionary transition from "non-bilaterian" animals, which are those lacking a bilaterally symmetric body plan (Porifera, Ctenophora, Cnidaria and Placozoa), to "bilaterian" animals (Bilateria) whose body plans display bilateral symmetry. The latter are further classified based on a major division between Deuterostomes and Protostomes. The relationships among non-bilaterian animals are disputed, but all bilaterian animals are thought to form a monophyletic group. Current understanding of the relationships among the major groups of animals is summarized by the following cladogram:

Non-bilaterian Animals: Porifera, Placozoa, Ctenophora, Cnidaria

Several animal phyla are recognized for their lack of bilateral symmetry, and are thought to have diverged from other animals early in evolution. Among these, the sponges (Porifera) were long thought to have diverged first, representing the oldest animal phylum. They lack the complex organization found in most other phyla. Their cells are differentiated, but in most cases not organized into distinct tissues. Sponges typically feed by drawing in water through pores. However, a series of phylogenomic studies from 2008-2015 have found support for Ctenophora, or comb jellies, as the basal lineage of animals. This result has been controversial, since it would imply that sponges may not be so primitive, but may instead be secondarily simplified. Other researchers have argued that the placement of Ctenophora as the earliest-diverging animal phylum is a statistical anomaly caused by the high rate of evolution in ctenophore genomes.

Among the other phyla, the Ctenophora and the Cnidaria, which includes sea anemones, corals, and jellyfish, are radially symmetric and have digestive chambers with a single opening, which serves as both the mouth and the anus. Both have distinct tissues, but they are not organized into organs. There are only two main germ layers, the ectoderm and endoderm, with only scattered cells between them. As such, these animals are sometimes called diploblastic. The tiny placozoans are similar, but they do not have a permanent digestive chamber.

The Myxozoa, microscopic parasites that were originally considered Protozoa, are now believed to have evolved within Cnidaria.

Orange elephant ear sponge, *Agelas clathrodes*, in foreground. Two corals in the background: a sea fan, *Iciligorgia schrammi*, and a sea rod, *Plexaurella nutans*.

Bilaterian Animals

The remaining animals form a monophyletic group called the Bilateria. For the most part, they are bilaterally symmetric, and often have a specialized head with feeding and sensory organs. The body is triploblastic, i.e. all three germ layers are well-developed, and tissues form distinct organs. The digestive chamber has two openings, a mouth and an anus, and there is also an internal body cavity called a coelom or pseudocoelom. There are exceptions to each of these characteristics, however—for instance adult echinoderms are radially symmetric, and certain parasitic worms have extremely simplified body structures.

Genetic studies have considerably changed our understanding of the relationships within the Bilateria. Most appear to belong to two major lineages: the deuterostomes and the protostomes, the latter of which includes the Ecdysozoa, and Lophotrochozoa. The Chaetognatha or arrow worms have been traditionally classified as deuterostomes, though recent molecular studies have identified this group as a basal protostome lineage.

In addition, there are a few small groups of bilaterians with relatively cryptic morphology whose relationships with other animals are not well-established. For example, recent molecular studies have identified Acoelomorpha and *Xenoturbella* as comprising a monophyletic group, but studies disagree as to whether this group evolved from within deuterostomes, or whether it represents the sister group to all other bilaterian animals (Nephrozoa). Other groups of uncertain affinity include the Rhombozoa and Orthonectida. One phyla, the Monoblastozoa, was described by a scientist in 1892, but so far there have been no evidence of its existence.

Deuterostomes and Protostomes

Superb fairy-wren, *Malurus cyaneus*

Deuterostomes differ from protostomes in several ways. Animals from both groups possess a complete digestive tract. However, in protostomes, the first opening of the gut to appear in embryological development (the archenteron) develops into the mouth, with the anus forming secondarily. In deuterostomes the anus forms first, with the mouth developing secondarily. In most protostomes, cells simply fill in the interior of the gastrula to form the mesoderm, called schizocoelous development, but in deuterostomes, it forms through invagination of the endoderm, called enterocoelic pouching. Deuterostome embryos undergo radial cleavage during cell division, while protostomes undergo spiral cleavage.

All this suggests the deuterostomes and protostomes are separate, monophyletic lineages. The main phyla of deuterostomes are the Echinodermata and Chordata. The former are radially symmetric and exclusively marine, such as starfish, sea urchins, and sea cucumbers. The latter are dominated by the vertebrates, animals with backbones. These include fish, amphibians, reptiles, birds, and mammals.

In addition to these, the deuterostomes also include the Hemichordata, or acorn worms, which are thought to be closely related to Echinodermata forming a group known as Ambulacraria. Although they are not especially prominent today, the important fossil graptolites may belong to this group.

Ecdysozoa

Yellow-winged darter, *Sympetrum flaveolum*

The Ecdysozoa are protostomes, named after the common trait of growth by moulting or ecdysis. The largest animal phylum belongs here, the Arthropoda, including insects, spiders, crabs, and their kin. All these organisms have a body divided into repeating segments, typically with paired appendages. Two smaller phyla, the Onychophora and Tardigrada, are close relatives of the arthropods and share these traits. The ecdysozoans also include the Nematoda or roundworms, perhaps the second largest animal phylum. Roundworms are typically microscopic, and occur in nearly every environment where there is water. A number are important parasites. Smaller phyla related to them are the Nematomorpha or horsehair worms, and the Kinorhyncha, Priapulida, and Loricifera. These groups have a reduced coelom, called a pseudocoelom.

Roman snail, *Helix pomatia*

Lophotrochozoa

The Lophotrochozoa, evolved within Protostomia, include two of the most successful animal phyla, the Mollusca and Annelida. The former, which is the second-largest animal phylum by number of described species, includes animals such as snails, clams, and squids, and the latter comprises the segmented worms, such as earthworms and leeches. These two groups have long been considered close relatives because of the common presence of trochophore larvae, but the annelids were considered closer to the arthropods because they are both segmented. Now, this is generally considered convergent evolution, owing to many morphological and genetic differences between the two phyla. Lophotrochozoa also includes the Nemertea or ribbon worms, the Sipuncula, and several phyla that have a ring of ciliated tentacles around the mouth, called a lophophore. These were traditionally grouped together as the lophophorates. but it now appears that the lophophorate group may be paraphyletic, with some closer to the nemerteans and some to the molluscs and annelids. They include the Brachiopoda or lamp shells, which are prominent in the fossil record, the Entoprocta, the Phoronida, and possibly the Bryozoa or moss animals.

The Platyzoa include the phylum Platyhelminthes, the flatworms. These were originally considered some of the most primitive Bilateria, but it now appears they developed from more complex ancestors. A number of parasites are included in this group, such as the flukes and tapeworms. Flatworms are acoelomates, lacking a body cavity, as are their closest relatives, the microscopic Gastrotricha. The other platyzoan phyla are mostly microscopic and pseudocoelomate. The most prominent are the Rotifera or rotifers, which are common in aqueous environments. They also include the Acanthocephala or spiny-headed worms, the Gnathostomulida, Micrognathozoa, and possibly the Cycliophora. These groups share the presence of complex jaws, from which they are called the Gnathifera.

A relationship between the Brachiopoda and Nemertea has been suggested by molecular data. A second study has also suggested this relationship. This latter study also suggested that Annelida and Mollusca may be sister clades. Another study has suggested that Annelida and Mollusca are sister clades. This clade has been termed the Neotrochozoa.

Number of Extant Species

Animals can be divided into two broad groups: vertebrates (animals with a backbone) and invertebrates (animals without a backbone). Half of all described vertebrate species are fishes and three-quarters of all described invertebrate species are insects. The following table lists the number of described extant species for each major animal subgroup as estimated for the IUCN Red List of Threatened Species, *2014.3*.

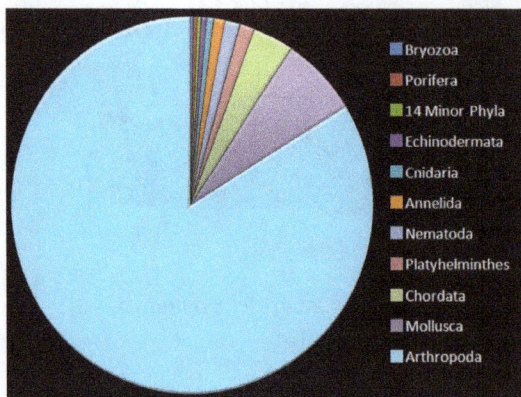

The relative number of species contributed to the total by each phylum of animals

Group	Image	Subgroup	Estimated number of described species
Vertebrates		Fishes	32,900
		Amphibians	7,302
		Reptiles	10,038
		Birds	10,425
		Mammals	5,513
Total vertebrate species: 66,178			

		Insects	1,000,000
Invertebrates		Insects	1,000,000
		Molluscs	85,000
		Crustaceans	47,000
		Corals	2,000
		Arachnids	102,248
		Velvet worms	165
		Horseshoe crabs	4
		Others	68,658
Total invertebrate species: 1,305,075			
Total for all animal species: 1,371,253			

Over 95% of the described animal species in the world are invertebrates.

Model Organisms

Because of the great diversity found in animals, it is more economical for scientists to study a small number of chosen species so that connections can be drawn from their work and conclusions extrapolated about how animals function in general. Because they are easy to keep and breed, the fruit fly *Drosophila melanogaster* and the nematode *Caenorhabditis elegans* have long been the most intensively studied metazoan model organisms, and were among the first life-forms to be genetically sequenced. This was facilitated by the severely reduced state of their genomes, but as many genes, introns, and linkages lost, these ecdysozoans can teach us little about the origins of animals in general. The extent of this type of evolution within the superphylum will be revealed by the crustacean, annelid, and molluscan genome projects currently in progress. Analysis of the starlet sea anemone genome has emphasized the importance of sponges, placozoans, and choanoflagellates, also being sequenced, in explaining the arrival of 1500 ancestral genes unique to the Eumetazoa.

An analysis of the homoscleromorph sponge *Oscarella carmela* also suggests that the last common ancestor of sponges and the eumetazoan animals was more complex than previously assumed.

Other model organisms belonging to the animal kingdom include the house mouse (*Mus musculus*), laboratory rat (*Rattus norvegicus*) and zebrafish (*Danio rerio*).

Animal Science

Animal Science (also Animal Bioscience) is described as "studying the biology of animals that are under the control of humankind". Historically, the degree was called animal husbandry and the animals studied were livestock species, like cattle, sheep, pigs, poultry, and horses. Today, courses available now look at a far broader area to include companion animals like dogs and cats, and many exotic species. Degrees in Animal Science are offered at a number of colleges and universities. In the United States, the universities offering such a program were Land Grant Universities and include University of Nebraska–Lincoln, Cornell University, UC Davis, Michigan State University, Purdue University, The Ohio State University, The Pennsylvania State University, Iowa State University and the University of Minnesota. Typically, the Animal Science curriculum not only provides a strong science background, but also hands-on experience working with animals on campus-based farms.

Education

Professional education in animal science prepares students for career opportunities in areas such as animal breeding, food and fiber production, nutrition, animal agribusiness, animal behavior and welfare. Courses in a typical Animal Science program may include genetics, microbiology, animal behavior, nutrition, physiology, and reproduction. Courses in support areas, such as genetics, soils, agricultural economics and marketing, legal aspects, and the environment also are offered. All of these courses are essential to entering an animal science profession.

Bachelor Degree

At many universities, a Bachelor of Science (BS) degree in Animal Science allows emphasis in certain areas. Typical areas are species-specific or career-specific. Species-specific areas of emphasis prepare students for a career in dairy management, beef management, swine management, sheep or small ruminant management, poultry production, or the horse industry. Other career-specific areas of study include pre-veterinary medicine studies, livestock business and marketing, animal welfare and behavior, animal nutrition science, animal reproduction science, or genetics. Youth programs are also an important part of animal science programs.

Pre-veterinary Emphasis

Many schools that offer a degree option in Animal Science also offer a pre-veterinary emphasis such as Iowa State University, the University of Nebraska-Lincoln and the University of Minnesota, for example. This option provides an in-depth knowledge base of the biological and physical sciences including nutrition, reproduction, physiology, and genetics. This can prepare students for graduate studies in animal science, veterinary school, and pharmaceutical or animal science industries.

Graduate Studies

In a Master of Science degree option, students take required courses in areas that support their main interest. These courses are above courses normally required for a Bachelor of Science degree in the Animal Science major. For example, in a Ph.D. degree program students take courses related to their major that are more in depth than those for the Master of Science degree, with an emphasis on research or teaching.

Graduate studies in animal sciences are considered preparation for upper level positions in production, management, education, research, or agriservices. Professional study in veterinary medicine, law, and business administration are among the most commonly chosen programs by graduates. Other areas of study include growth biology, physiology, nutrition, and production systems.

References

- Davidson, Michael W. "Animal Cell Structure". Archived from the original on 20 September 2007. Retrieved 20 September 2007

- Calisher, CH (2007). "Taxonomy: what's in a name? Doesn't a rose by any other name smell as sweet?". Croatian Medical Journal. 48 (2): 268–270. PMC 2080517. PMID 17436393

- Alberts, Bruce; Johnson, Alexander; Lewis, Julian; Raff, Martin; Roberts, Keith; Walter, Peter (2002). Molecular Biology of the Cell (4th ed.). New York: Garland Science. Retrieved 2015-03-23

- Ralls K, Ballou J (April 1982). "Effect of inbreeding on juvenile mortality in some small mammal species". Lab. Anim. 16 (2): 159–66. PMID 7043080. doi:10.1258/002367782781110151

- Todaro, Antonio. "Gastrotricha: Overview". Gastrotricha: World Portal. University of Modena & Reggio Emilia. Retrieved 2008-01-26

- Kim, Chang Bae; Moon, Seung Yeo; Gelder, Stuart R.; Kim, Won (September 1996). "Phylogenetic Relationships of Annelids, Molluscs, and Arthropods Evidenced from Molecules and Morphology". Journal of Molecular Evolution. New York: Springer. 43 (3): 207–215. PMID 8703086. doi:10.1007/PL00006079

- "Biodiversity: Mollusca". The Scottish Association for Marine Science. Archived from the original on 8 July 2006. Retrieved 2007-11-19

- Pusey A, Wolf M (1996). "Inbreeding avoidance in animals". Trends Ecol. Evol. (Amst.). 11 (5): 201–6. PMID 21237809. doi:10.1016/0169-5347(96)10028-8

Classification of Animals

Animals can be classified into parazoa and eumetazoa. Parazoa consists of sponges and the Trichoplax adhaerens (Placozoa) whereas eumetazoa consists of radiate, bilateria, mesozoa and nephrozoa. This chapter has been carefully written to provide an easy understanding of the classification of animals.

Parazoa

The Parazoa are an ancestral subkingdom of animals, literally translated as "beside the animals".

Description

Parazoans differ from their choanoflagellate ancestors in that they are not microscopic and have differentiated cells. However, they are an outgroup of the animal phylogenetic tree since they do not have tissues or organs. The only surviving parazoans are the sponges, which belong to the phylum Porifera, and the Trichoplax in the phylum Placozoa.

Parazoa display no body symmetry (are asymmetrical); all other animal groups display some sort of symmetry. There are currently 5000 species, 150 of which are freshwater. Larvae are planktonic and adults are sessile.

Cladistics

The Parazoa-Eumetazoa split has been estimated at 940 million years ago.

The parazoa group is now considered paraphyletic. It is not included in most modern cladistic analyses. When referenced, it is sometimes considered an equivalent to Porifera.

Some authors include Placozoa, a phylum that consists of only one species, *Trichoplax adhaerens*, in the division, but they are also sometimes placed in the subkingdom Agnotozoa.

Sponge

Sponges are the basalmost clade of animals of the phylum Porifera (meaning "pore bearer"). They are multicellular parazoan organisms that have bodies full of pores and channels allowing water to circulate through them, consisting of jelly-like mesohyl sandwiched between two thin layers of cells. Sponges have unspecialized cells that can transform into other types and that often migrate between the main cell layers and the mesohyl in the process. Sponges do not have nervous, digestive or circulatory systems. Instead, most rely on maintaining a constant water flow through their bodies to obtain food and oxygen and to remove wastes.

Overview

Sponge biodiversity and morphotypes at the lip of a wall site in 60 feet (20 m) of water. Included are the yellow tube sponge, *Aplysina fistularis*, the purple vase sponge, *Niphates digitalis*, the red encrusting sponge, *Spiratrella coccinea*, and the gray rope sponge, *Callyspongia* sp.

Sponges are similar to other animals in that they are multicellular, heterotrophic, lack cell walls and produce sperm cells. Unlike other animals, they lack true tissues and organs, and have no body symmetry. The shapes of their bodies are adapted for maximal efficiency of water flow through the central cavity, where it deposits nutrients, and leaves through a hole called the osculum. Many sponges have internal skeletons of spongin and/or spicules of calcium carbonate or silicon dioxide. All sponges are sessile aquatic animals. Although there are freshwater species, the great majority are marine (salt water) species, ranging from tidal zones to depths exceeding 8,800 m (5.5 mi).

While most of the approximately 5,000–10,000 known species feed on bacteria and other food particles in the water, some host photosynthesizing micro-organisms as endosymbionts and these alliances often produce more food and oxygen than they consume. A few species of sponge that live in food-poor environments have become carnivores that prey mainly on small crustaceans.

Most species use sexual reproduction, releasing sperm cells into the water to fertilize ova that in some species are released and in others are retained by the "mother". The fertilized eggs form larvae which swim off in search of places to settle. Sponges are known for regenerating from fragments that are broken off, although this only works if the fragments include the right types of cells. A few species reproduce by budding. When conditions deteriorate, for example as temperatures drop, many freshwater species and a few marine ones produce gemmules, "survival pods" of unspecialized cells that remain dormant until conditions improve and then either form completely new sponges or recolonize the skeletons of their parents.

The mesohyl functions as an endoskeleton in most sponges, and is the only skeleton in soft sponges that encrust hard surfaces such as rocks. More commonly, the mesohyl is stiffened by mineral spicules, by spongin fibers or both. Demosponges use spongin, and in many species, silica spicules and in some species, calcium carbonate exoskeletons. Demosponges constitute about 90% of all known sponge species, including all freshwater ones, and have the widest range of habitats. Calcareous sponges, which have calcium carbonate spicules and, in some species, calcium carbonate exoskeletons, are restricted to relatively shallow marine waters where production of calcium carbonate is easiest. The fragile glass sponges, with "scaffolding" of silica spicules, are restricted to polar regions and the ocean depths where predators are rare. Fossils of all of these types have been found in rocks dated from 580 million years ago. In addition Archaeocyathids, whose fossils are common in rocks from 530 to 490 million years ago, are now regarded as a type of sponge.

The single-celled choanoflagellates resemble the choanocyte cells of sponges which are used to drive their water flow systems and capture most of their food. This along with phylogenetic studies of ribosomal molecules have been used as morphological evidence to suggest sponges are the sister group to the rest of animals. Some studies have shown that sponges do not form a monophyletic group, in other words do not include *all and only* the descendants of a common ancestor. Recent phylogenetic analyses suggest that comb jellies rather than sponges are the sister group to the rest of animals.

The few species of demosponge that have entirely soft fibrous skeletons with no hard elements have been used by humans over thousands of years for several purposes, including as padding and as cleaning tools. By the 1950s, though, these had been overfished so heavily that the industry almost collapsed, and most sponge-like materials are now synthetic. Sponges and their microscopic endosymbionts are now being researched as possible sources of medicines for treating a wide range of diseases. Dolphins have been observed using sponges as tools while foraging.

Distinguishing Features

Sponges constitute the phylum Porifera, and have been defined as sessile metazoans (multicelled immobile animals) that have water intake and outlet openings connected by chambers lined with choanocytes, cells with whip-like flagella. However, a few carnivorous sponges have lost these water flow systems and the choanocytes. All known living sponges can remold their bodies, as most types of their cells can move within their bodies and a few can change from one type to another.

Like cnidarians (jellyfish, etc.) and ctenophores (comb jellies), and unlike all other known metazoans, sponges' bodies consist of a non-living jelly-like mass (mesoglea) sandwiched between two main layers of cells. Cnidarians and ctenophores have simple nervous systems, and their cell layers are bound by internal connections and by being mounted on a basement membrane (thin fibrous mat, also known as "basal lamina"). Sponges have no nervous systems, their middle jelly-like layers have large and varied populations of cells, and some types of cells in their outer layers may move into the middle layer and change their functions.

	Sponges	**Cnidarians and ctenophores**
Nervous system	No	Yes, simple
Cells in each layer bound together	No, except that Homoscleromorpha have basement membranes.	Yes: inter-cell connections; basement membranes
Number of cells in middle "jelly" layer	Many	Few
Cells in outer layers can move inwards and change functions	Yes	No

Basic Structure

Cell Types

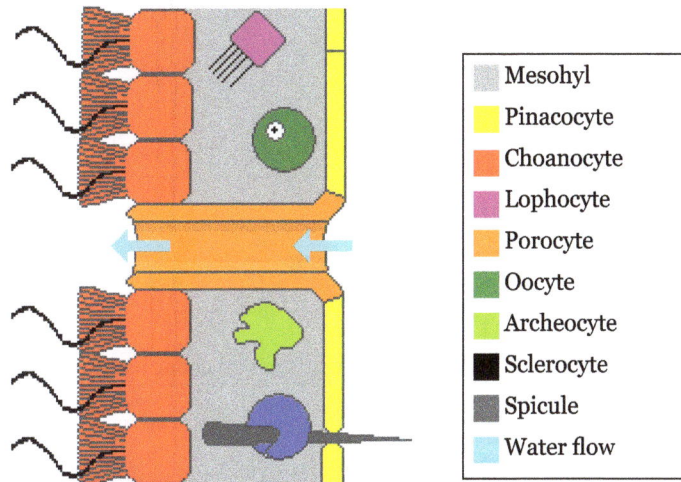

Main cell types of Porifera

A sponge's body is hollow and is held in shape by the mesohyl, a jelly-like substance made mainly of collagen and reinforced by a dense network of fibers also made of collagen. The inner surface is covered with choanocytes, cells with cylindrical or conical collars surrounding one flagellum per choanocyte. The wave-like motion of the whip-like flagella drives water through the sponge's body. All sponges have ostia, channels leading to the interior through the mesohyl, and in most sponges these are controlled by tube-like porocytes that form closable inlet valves. Pinacocytes, plate-like cells, form a single-layered external skin over all other parts of the mesohyl that are not covered by choanocytes, and the pinacocytes also digest food particles that are too large to enter the ostia, while those at the base of the animal are responsible for anchoring it.

Other types of cell live and move within the mesohyl:

- Lophocytes are amoeba-like cells that move slowly through the mesohyl and secrete collagen fibres.

- Collencytes are another type of collagen-producing cell.

- Rhabdiferous cells secrete polysaccharides that also form part of the mesohyl.

- Oocytes and spermatocytes are reproductive cells.

- Sclerocytes secrete the mineralized spicules ("little spines") that form the skeletons of many sponges and in some species provide some defense against predators.

- In addition to or instead of sclerocytes, demosponges have spongocytes that secrete a form of collagen that polymerizes into spongin, a thick fibrous material that stiffens the mesohyl.

- Myocytes ("muscle cells") conduct signals and cause parts of the animal to contract.

- "Grey cells" act as sponges' equivalent of an immune system.

- Archaeocytes (or amoebocytes) are amoeba-like cells that are totipotent, in other words each is capable of transformation into any other type of cell. They also have important roles in feeding and in clearing debris that block the ostia.

Glass Sponges' Syncytia

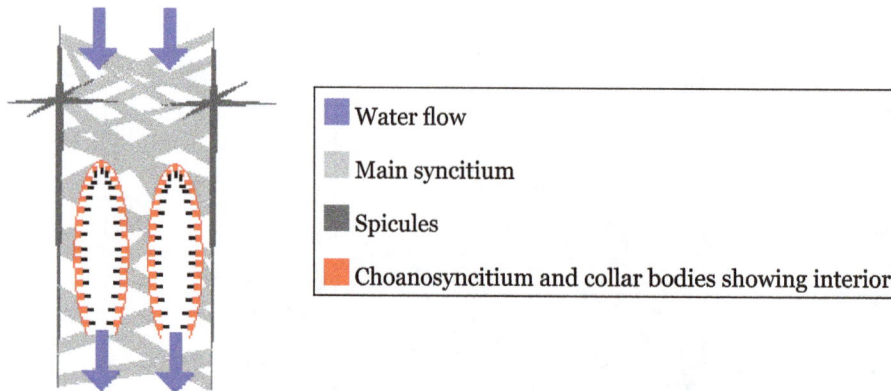

■	Water flow
■	Main syncitium
■	Spicules
■	Choanosyncitium and collar bodies showing interior

The glass sponge *Euplectella*

Glass sponges present a distinctive variation on this basic plan. Their spicules, which are made of silica, form a scaffolding-like framework between whose rods the living tissue is suspended like a cobweb that contains most of the cell types. This tissue is a syncytium that in some ways behaves like many cells that share a single external membrane, and in others like a single cell with multiple nuclei. The mesohyl is absent or minimal. The syncytium's cytoplasm, the soupy fluid that fills the interiors of cells, is organized into "rivers" that transport nuclei, organelles ("organs" within cells) and other substances. Instead of choanocytes, they have further syncytia, known as choanosyn-cytia, which form bell-shaped chambers where water enters via perforations. The insides of these chambers are lined with "collar bodies", each consisting of a collar and flagellum but without a nucleus of its own. The motion of the flagella sucks water through passages in the "cobweb" and expels it via the open ends of the bell-shaped chambers.

Some types of cells have a single nucleus and membrane each, but are connected to other sin-gle-nucleus cells and to the main syncytium by "bridges" made of cytoplasm. The sclerocytes that build spicules have multiple nuclei, and in glass sponge larvae they are connected to other tissues by cytoplasm bridges; such connections between sclerocytes have not so far been found in adults, but this may simply reflect the difficulty of investigating such small-scale features. The bridges are controlled by "plugged junctions" that apparently permit some substances to pass while blocking others.

Water Flow and Body Structures

Most sponges work rather like chimneys: they take in water at the bottom and eject it from the osculum ("little mouth") at the top. Since ambient currents are faster at the top, the suction effect that they produce by Bernoulli's principle does some of the work for free. Sponges can control the water flow by various combinations of wholly or partially closing the osculum and ostia (the intake pores) and varying the beat of the flagella, and may shut it down if there is a lot of sand or silt in the water.

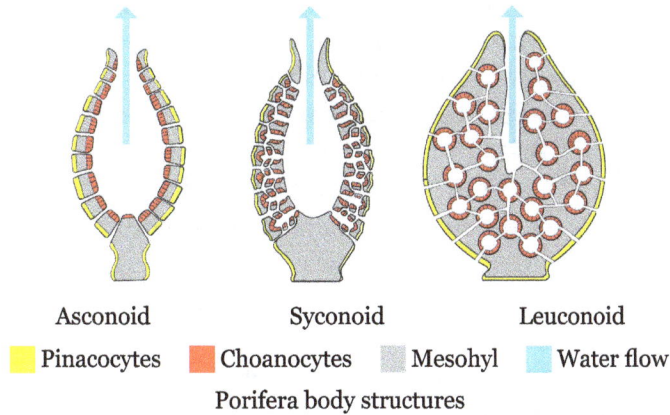

Asconoid   Syconoid   Leuconoid

Pinacocytes   Choanocytes   Mesohyl   Water flow

Porifera body structures

Although the layers of pinacocytes and choanocytes resemble the epithelia of more complex animals, they are not bound tightly by cell-to-cell connections or a basal lamina (thin fibrous sheet underneath). The flexibility of these layers and re-modeling of the mesohyl by lophocytes allow the animals to adjust their shapes throughout their lives to take maximum advantage of local water currents.

The simplest body structure in sponges is a tube or vase shape known as "asconoid", but this severely limits the size of the animal. The body structure is characterized by a stalk-like spongocoel surrounded by a single layer of choanocytes. If it is simply scaled up, the ratio of its volume to surface area increases, because surface increases as the square of length or width while volume increases proportionally to the cube. The amount of tissue that needs food and oxygen is determined by the volume, but the pumping capacity that supplies food and oxygen depends on the area covered by choanocytes. Asconoid sponges seldom exceed 1 mm (0.039 in) in diameter.

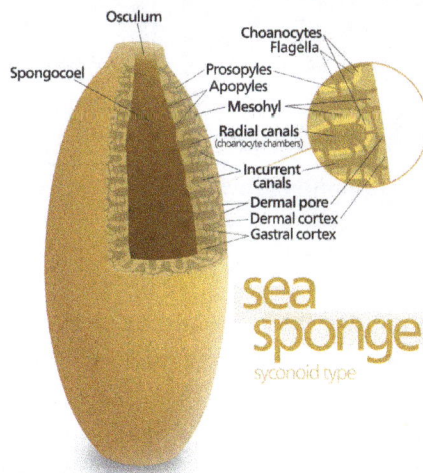

Diagram of a syconoid sponge.

Some sponges overcome this limitation by adopting the "syconoid" structure, in which the body wall is pleated. The inner pockets of the pleats are lined with choanocytes, which connect to the outer pockets of the pleats by ostia. This increase in the number of choanocytes and hence in pumping capacity enables syconoid sponges to grow up to a few centimeters in diameter.

The "leuconoid" pattern boosts pumping capacity further by filling the interior almost completely with mesohyl that contains a network of chambers lined with choanocytes and connected to each other and to the water intakes and outlet by tubes. Leuconid sponges grow to over 1 m (3.3 ft) in diameter, and the fact that growth in any direction increases the number of choanocyte chambers enables them to take a wider range of forms, for example "encrusting" sponges whose shapes follow those of the surfaces to which they attach. All freshwater and most shallow-water marine sponges have leuconid bodies. The networks of water passages in glass sponges are similar to the leuconid structure. In all three types of structure the cross-section area of the choanocyte-lined regions is much greater than that of the intake and outlet channels. This makes the flow slower near the choanocytes and thus makes it easier for them to trap food particles. For example, in *Leuconia*, a small leuconoid sponge about 10 centimetres (3.9 in) tall and 1 centimetre (0.39 in) in diameter, water enters each of more than 80,000 intake canals at 6 cm per *minute*. However, because *Leuconia* has more than 2 million flagellated chambers whose combined diameter is much greater than that of the canals, water flow through chambers slows to 3.6 cm per *hour*, making it easy for choanocytes to capture food. All the water is expelled through a single osculum at about 8.5 cm per *second*, fast enough to carry waste products some distance away.

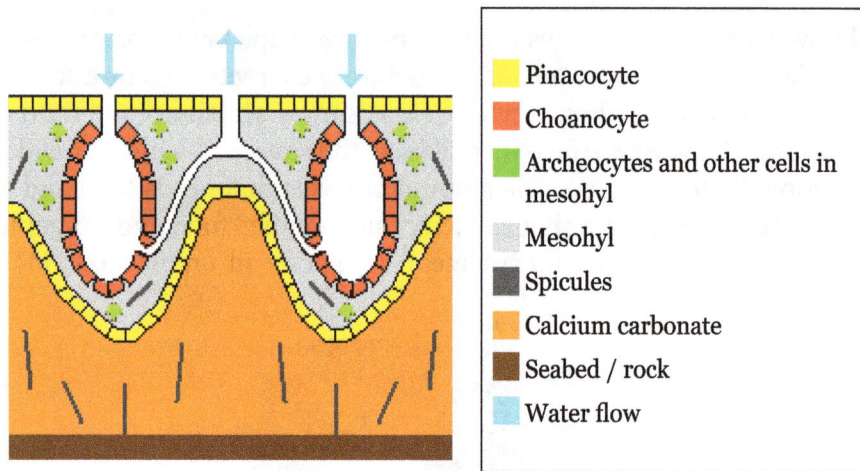

🟨	Pinacocyte
🟧	Choanocyte
🟩	Archeocytes and other cells in mesohyl
⬜	Mesohyl
⬛	Spicules
🟧	Calcium carbonate
🟫	Seabed / rock
🟦	Water flow

Sponge with calcium carbonate skeleton

Skeleton

In zoology a skeleton is any fairly rigid structure of an animal, irrespective of whether it has joints and irrespective of whether it is biomineralized. The mesohyl functions as an endoskeleton in most sponges, and is the only skeleton in soft sponges that encrust hard surfaces such as rocks. More commonly the mesohyl is stiffened by mineral spicules, by spongin fibers or both. Spicules may be made of silica or calcium carbonate, and vary in shape from simple rods to three-dimensional "stars" with up to six rays. Spicules are produced by sclerocyte cells, and may be separate, connected by joints, or fused.

Some sponges also secrete exoskeletons that lie completely outside their organic components. For example, sclerosponges ("hard sponges") have massive calcium carbonate exoskeletons over which the organic matter forms a thin layer with choanocyte chambers in pits in the mineral. These exoskeletons are secreted by the pinacocytes that form the animals' skins.

Vital Functions

Spongia officinalis, "the kitchen sponge", is dark grey when alive

Movement

Although adult sponges are fundamentally sessile animals, some marine and freshwater species can move across the sea bed at speeds of 1–4 mm (0.039–0.157 in) per day, as a result of amoeba-like movements of pinacocytes and other cells. A few species can contract their whole bodies, and many can close their oscula and ostia. Juveniles drift or swim freely, while adults are stationary.

Respiration, Feeding and Excretion

Sponges do not have distinct circulatory, respiratory, digestive, and excretory systems – instead the water flow system supports all these functions. They filter food particles out of the water flowing through them. Particles larger than 50 micrometers cannot enter the ostia and pinacocytes consume them by phagocytosis (engulfing and internal digestion). Particles from 0.5 μm to 50 μm are trapped in the ostia, which taper from the outer to inner ends. These particles are consumed by pinacocytes or by archaeocytes which partially extrude themselves through the walls of the ostia. Bacteria-sized particles, below 0.5 micrometers, pass through the ostia and are caught and consumed by choanocytes. Since the smallest particles are by far the most common, choanocytes typically capture 80% of a sponge's food supply. Archaeocytes transport food packaged in vesicles from cells that directly digest food to those that do not. At least one species of sponge has internal fibers that function as tracks for use by nutrient-carrying archaeocytes, and these tracks also move inert objects.

It used to be claimed that glass sponges could live on nutrients dissolved in sea water and were very averse to silt. However a study in 2007 found no evidence of this and concluded that they extract bacteria and other micro-organisms from water very efficiently (about 79%) and process suspended sediment grains to extract such prey. Collar bodies digest food and distribute it wrapped in vesicles that are transported by dynein "motor" molecules along bundles of microtubules that run throughout the syncytium.

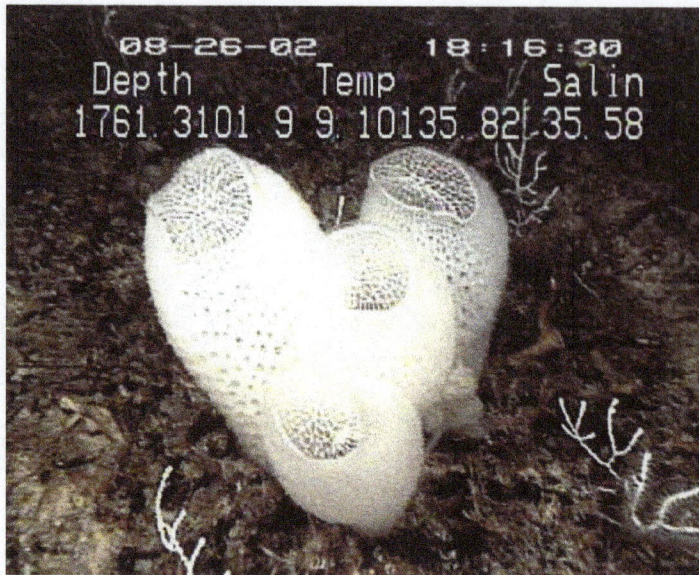

Euplectella aspergillum, a glass sponge known as "Venus' Flower Basket"

Sponges' cells absorb oxygen by diffusion from water into cells as water flows through body, into which carbon dioxide and other soluble waste products such as ammonia also diffuse. Archeocytes remove mineral particles that threaten to block the ostia, transport them through the mesohyl and generally dump them into the outgoing water current, although some species incorporate them into their skeletons.

Carnivorous Sponges

A few species that live in waters where the supply of food particles is very poor prey on crustaceans and other small animals. So far only 137 species have been discovered. Most belong to the family Cladorhizidae, but a few members of the Guitarridae and Esperiopsidae are also carnivores. In most cases little is known about how they actually capture prey, although some species are thought to use either sticky threads or hooked spicules. Most carnivorous sponges live in deep waters, up to 8,840 m (5.49 mi), and the development of deep-ocean exploration techniques is expected to lead to the discovery of several more. However one species has been found in Mediterranean caves at depths of 17–23 m (56–75 ft), alongside the more usual filter feeding sponges. The cave-dwelling predators capture crustaceans under 1 mm (0.039 in) long by entangling them with fine threads, digest them by enveloping them with further threads over the course of a few days, and then return to their normal shape; there is no evidence that they use venom.

Most known carnivorous sponges have completely lost the water flow system and choanocytes. However the genus *Chondrocladia* uses a highly modified water flow system to inflate balloon-like structures that are used for capturing prey.

Endosymbionts

Freshwater sponges often host green algae as endosymbionts within archaeocytes and other cells, and benefit from nutrients produced by the algae. Many marine species host other photosynthesizing organisms, most commonly cyanobacteria but in some cases dinoflagellates. Symbiotic cyanobacteria may form a third of the total mass of living tissue in some sponges, and some sponges gain 48% to 80% of their energy supply from these micro-organisms. In 2008 a University of Stuttgart team reported that spicules made of silica conduct light into the mesohyl, where the photosynthesizing endosymbionts live. Sponges that host photosynthesizing organisms are most common in waters with relatively poor supplies of food particles, and often have leafy shapes that maximize the amount of sunlight they collect.

A recently discovered carnivorous sponge that lives near hydrothermal vents hosts methane-eating bacteria, and digests some of them.

"Immune" System

Sponges do not have the complex immune systems of most other animals. However they reject grafts from other species but accept them from other members of their own species. In a few marine species, gray cells play the leading role in rejection of foreign material. When invaded, they produce a chemical that stops movement of other cells in the affected area, thus preventing the intruder from using the sponge's internal transport systems. If the intrusion persists, the grey cells concentrate in the area and release toxins that kill all cells in the area. The "immune" system can stay in this activated state for up to three weeks.

Reproduction

Asexual

The freshwater sponge *Spongilla lacustris*

Sponges have three asexual methods of reproduction: after fragmentation; by budding; and by producing gemmules. Fragments of sponges may be detached by currents or waves. They use the mobility of their pinacocytes and choanocytes and reshaping of the mesohyl to re-attach themselves to a suitable surface and then rebuild themselves as small but functional sponges over the course of several days. The same capabilities enable sponges that have been squeezed through a fine cloth to regenerate. A sponge fragment can only regenerate if it contains both collencytes to produce mesohyl and archeocytes to produce all the other cell types. A very few species reproduce by budding.

Gemmules are "survival pods" which a few marine sponges and many freshwater species produce by the thousands when dying and which some, mainly freshwater species, regularly produce in autumn. Spongocytes make gemmules by wrapping shells of spongin, often reinforced with spicules, round clusters of archeocytes that are full of nutrients. Freshwater gemmules may also include phytosynthesizing symbionts. The gemmules then become dormant, and in this state can survive cold, drying out, lack of oxygen and extreme variations in salinity. Freshwater gemmules often do not revive until the temperature drops, stays cold for a few months and then reaches a "near-normal" level. When a gemmule germinates, the archeocytes round the outside of the cluster transform into pinacocytes, a membrane over a pore in the shell bursts, the cluster of cells slowly emerges, and most of the remaining archeocytes transform into other cell types needed to make a functioning sponge. Gemmules from the same species but different individuals can join forces to form one sponge. Some gemmules are retained within the parent sponge, and in spring it can be difficult to tell whether an old sponge has revived or been "recolonized" by its own gemmules.

Sexual

Most sponges are hermaphrodites (function as both sexes simultaneously), although sponges have no gonads (reproductive organs). Sperm are produced by choanocytes or entire choanocyte chambers that sink into the mesohyl and form spermatic cysts while eggs are formed by transformation of archeocytes, or of choanocytes in some species. Each egg generally acquires a yolk by consuming "nurse cells". During spawning, sperm burst out of their cysts and are expelled via the osculum. If they contact another sponge of the same species, the water flow carries them to choanocytes that engulf them but, instead of digesting them, metamorphose to an ameboid form and carry the sperm through the mesohyl to eggs, which in most cases engulf the carrier and its cargo.

A few species release fertilized eggs into the water, but most retain the eggs until they hatch. There are four types of larvae, but all are balls of cells with an outer layer of cells whose flagellae or cilia enable the larvae to move. After swimming for a few days the larvae sink and crawl until they find a place to settle. Most of the cells transform into archeocytes and then into the types appropriate for their locations in a miniature adult sponge.

Glass sponge embryos start by dividing into separate cells, but once 32 cells have formed they rapidly transform into larvae that externally are ovoid with a band of cilia round the middle that they use for movement, but internally have the typical glass sponge structure of spicules with a cobweb-like main syncitium draped around and between them and choanosyncytia with multiple collar bodies in the center. The larvae then leave their parents' bodies.

Life Cycle

Sponges in temperate regions live for at most a few years, but some tropical species and perhaps some deep-ocean ones may live for 200 years or more. Some calcified demosponges grow by only 0.2 mm (0.0079 in) per year and, if that rate is constant, specimens 1 m (3.3 ft) wide must be about 5,000 years old. Some sponges start sexual reproduction when only a few weeks old, while others wait until they are several years old.

Coordination of Activities

Adult sponges lack neurons or any other kind of nervous tissue. However most species have the ability to perform movements that are coordinated all over their bodies, mainly contractions of the pinacocytes, squeezing the water channels and thus expelling excess sediment and other substances that may cause blockages. Some species can contract the osculum independently of the rest of the body. Sponges may also contract in order to reduce the area that is vulnerable to attack by predators. In cases where two sponges are fused, for example if there is a large but still unseparated bud, these contraction waves slowly become coordinated in both of the "Siamese twins". The coordinating mechanism is unknown, but may involve chemicals similar to neurotransmitters. However glass sponges rapidly transmit electrical impulses through all parts of the syncytium, and use this to halt the motion of their flagella if the incoming water contains toxins or excessive sediment. Myocytes are thought to be responsible for closing the osculum and for transmitting signals between different parts of the body.

Sponges contain genes very similar to those that contain the "recipe" for the post-synaptic density, an important signal-receiving structure in the neurons of all other animals. However, in sponges these genes are only activated in "flask cells" that appear only in larvae and may provide some sensory capability while the larvae are swimming. This raises questions about whether flask cells represent the predecessors of true neurons or are evidence that sponges' ancestors had true neurons but lost them as they adapted to a sessile lifestyle.

Ecology

Euplectella aspergillum is a deep ocean glass sponge; seen here at a depth of 2,572 metres (8,438 ft) off the coast of California.

Habitats

Sponges are worldwide in their distribution, living in a wide range of ocean habitats, from the polar regions to the tropics. Most live in quiet, clear waters, because sediment stirred up by waves or currents would block their pores, making it difficult for them to feed and breathe. The greatest numbers of sponges are usually found on firm surfaces such as rocks, but some sponges can attach themselves to soft sediment by means of a root-like base.

Sponges are more abundant but less diverse in temperate waters than in tropical waters, possibly because organisms that prey on sponges are more abundant in tropical waters. Glass sponges are the most common in polar waters and in the depths of temperate and tropical seas, as their very porous construction enables them to extract food from these resource-poor waters with the minimum of effort. Demosponges and calcareous sponges are abundant and diverse in shallower non-polar waters.

The different classes of sponge live in different ranges of habitat:

	Water type	Depth	Type of surface
Calcarea	Marine	less than 100 m (330 ft)	Hard
Glass sponges	Marine	Deep	Soft or firm sediment
Demosponges	Marine, brackish; and about 150 freshwater species	Inter-tidal to abyssal; a carnivorous demosponge has been found at 8,840 m (5.49 mi)	Any

As Primary Producers

Sponges with photosynthesizing endosymbionts produce up to three times more oxygen than they consume, as well as more organic matter than they consume. Such contributions to their habitats' resources are significant along Australia's Great Barrier Reef but relatively minor in the Caribbean.

Defenses

Holes made by clionaid sponge (producing the trace *Entobia*) after the death of a modern bivalve shell of species *Mercenaria mercenaria*, from North Carolina

Many sponges shed Sponge spicules, forming a dense carpet several meters deep that keeps away echinoderms which would otherwise prey on the sponges. They also produce toxins that prevent other sessile organisms such as bryozoans or sea squirts from growing on or near them, making sponges very effective competitors for living space. One of many examples includes ageliferin.

A few species, the Caribbean fire sponge *Tedania ignis*, cause a severe rash in humans who handle them. Turtles and some fish feed mainly on sponges. It is often said that sponges produce chemical defenses against such predators. However an experiment showed that there is no relationship between the toxicity of chemicals produced by sponges and how they taste to fish, which would diminish the usefulness of chemical defenses as deterrents. Predation by fish may even help to spread sponges by detaching fragments.

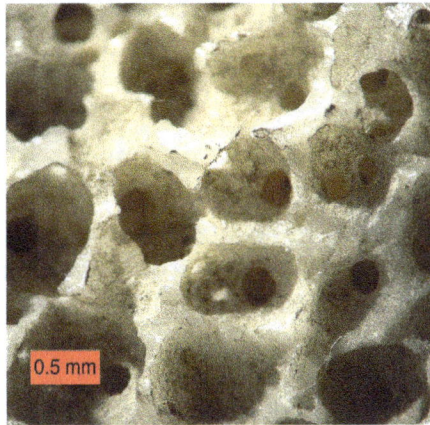

Close-up of the sponge boring *Entobia* in a modern oyster valve. Note the chambers which are connected by short tunnels.

Glass sponges produce no toxic chemicals, and live in very deep water where predators are rare.

Predation

Sponge flies, also known as spongilla-flies (Neuroptera, Sisyridae), are specialist predators of freshwater sponges. The female lays her eggs on vegetation overhanging water. The larvae hatch and drop into the water where they seek out sponges to feed on. They use their elongated mouthparts to pierce the sponge and suck the fluids within. The larvae of some species cling to the surface of the sponge while others take refuge in the sponge's internal cavities. The fully grown larvae leave the water and spin a cocoon in which to pupate.

Bioerosion

The Caribbean chicken-liver sponge *Chondrilla nucula* secretes toxins that kill coral polyps, allowing the sponges to grow over the coral skeletons. Others, especially in the family Clionaidae, use corrosive substances secreted by their archeocytes to tunnel into rocks, corals and the shells of dead mollusks. Sponges may remove up to 1 m (3.3 ft) per year from reefs, creating visible notches just below low-tide level.

Diseases

Caribbean sponges of the genus *Aplysina* suffer from Aplysina red band syndrome. This causes *Aplysina* to develop one or more rust-colored bands, sometimes with adjacent bands of necrotic tissue. These lesions may completely encircle branches of the sponge. The disease appears to be contagious and impacts approximately 10 percent of *A. cauliformis* on Bahamian reefs. The rust-colored bands are caused by a cyanobacterium, but it is unknown whether this organism actually causes the disease.

Collaboration with other Organisms

In addition to hosting photosynthesizing endosymbionts, sponges are noted for their wide range of collaborations with other organisms. The relatively large encrusting sponge *Lissodendoryx colombiensis* is most common on rocky surfaces, but has extended its range into seagrass meadows by letting itself be surrounded or overgrown by seagrass sponges, which are distasteful to the lo-cal starfish and therefore protect *Lissodendoryx* against them; in return the seagrass sponges get higher positions away from the sea-floor sediment.

Shrimps of the genus *Synalpheus* form colonies in sponges, and each shrimp species inhabits a different sponge species, making *Synalpheus* one of the most diverse crustacean genera. Spe-cifically, Synalpheus regalis utilizes the sponge not only as a food source, but also as a defense against other shrimp and predators. As many as 16,000 individuals inhabit a single loggerhead sponge, feeding off the larger particles that collect on the sponge as it filters the ocean to feed itself.

Systematics and Evolutionary History

Taxonomy

Linnaeus, who classified most kinds of sessile animals as belonging to the order Zoophyta in the class Vermes, mistakenly identified the genus *Spongia* as plants in the order Algae. For a long time thereafter sponges were assigned to a separate subkingdom, Parazoa ("beside the animals"), sepa-rate from the Eumetazoa which formed the rest of the kingdom Animalia. They have been regarded as a paraphyletic phylum, from which the higher animals have evolved. Other research indicates Porifera is monophyletic.

The phylum Porifera is further divided into classes mainly according to the composition of their skeletons:

- Hexactinellida (glass sponges) have silicate spicules, the largest of which have six rays and may be individual or fused. The main components of their bodies are syncytia in which large numbers of cell share a single external membrane.

- Calcarea have skeletons made of calcite, a form of calcium carbonate, which may form sep-arate spicules or large masses. All the cells have a single nucleus and membrane.

- Most Demospongiae have silicate spicules or spongin fibers or both within their soft tissues. However a few also have massive external skeletons made of aragonite, another form of calcium carbonate. All the cells have a single nucleus and membrane.

- Archeocyatha are known only as fossils from the Cambrian period.

In the 1970s, sponges with massive calcium carbonate skeletons were assigned to a separate class, Sclerospongiae, otherwise known as "coralline sponges". However, in the 1980s it was found that these were all members of either the Calcarea or the Demospongiae.

So far scientific publications have identified about 9,000 poriferan species, of which: about 400 are glass sponges; about 500 are calcareous species; and the rest are demosponges. However some types of habitat, vertical rock and cave walls and galleries in rock and coral boulders, have been investigated very little, even in shallow seas.

Classes

Sponges were traditionally distributed in three classes: calcareous sponges (Calcarea), glass sponges (Hexactinellida) and demosponges (Demospongiae). However, studies have shown that the Homoscleromorpha, a group thought to belong to the Demospongiae, is actually phylogenetically well separated. Therefore, they have recently been recognized as the fourth class of sponges.

Sponges are divided into classes mainly according to the composition of their skeletons:

	Type of cells	Spicules	Spongin fibers	Massive exoskeleton	Body form
Calcarea	Single nucleus, single external membrane	Calcite May be individual or large masses	Never	Common. Made of calcite if present.	Asconoid, syconoid, leuconoid or solenoid
Hexactinellida	Mostly syncytia in all species	Silica May be individual or fused	Never	Never	Leuconoid
Demospongiae	Single nucleus, single external membrane	Silica	In many species	In some species. Made of aragonite if present.	Leuconoid
Homoscleromorpha	Single nucleus, single external membrane	Silica	In many species	Never	Sylleibid or leuconoid

Fossil Record

Raphidonema faringdonense, a fossil sponge from the Cretaceous of England.

24-isopropylcholestane is a stable derivative of 24-isopropylcholesterol, which is said to be produced by demosponges but not by eumetazoans ("true animals", i.e. cnidarians and bilaterians). Since choanoflagellates are thought to be animals' closest single-celled relatives, a team of scientists examined the biochemistry and genes of one choanoflagellate species. They concluded that this species could not produce 24-isopropylcholesterol but that investigation of a wider range of choanoflagellates would be necessary in order to prove that the fossil 24-isopropylcholestane could only have been produced by demosponges. Although a previous publication reported traces of the chemical 24-isopropylcholestane in ancient rocks dating to 1,800 million years ago, recent research using a much more accurately dated rock series has revealed that these biomarkers only appear before the end of the Marinoan glaciation approximately 635 million years ago, and that "Biomarker analysis has yet to reveal any convincing evidence for ancient sponges pre-dating the first globally extensive Neoproterozoic glacial episode (the Sturtian, ~713 million years ago in Oman)". Nevertheless, this 'sponge biomarker' could have other sources – such as marine algae — so may not constrain the origin of Porifera.

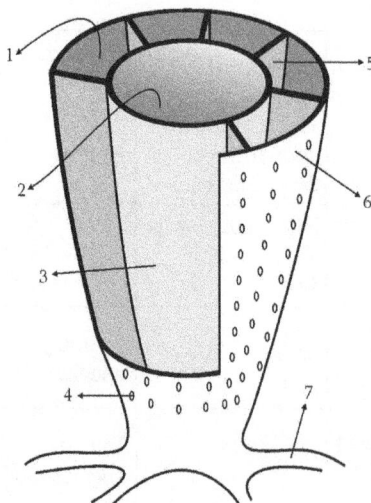

1: Gap **2:** Central cavity **3:** Internal wall **4:** Pore (all walls have pores) **5:** Septum **6:** Outer wall **7:** Holdfast
Archaeocyathid structure

Although molecular clocks and biomarkers suggest sponges existed well before the Cambrian explosion of life, silica spicules like those of demosponges are absent from the fossil record until the Cambrian, although one unsubstantiated report exists of spicules in rocks dated around 750 million years ago, although this appears unlikely based on the above reference. Well-preserved fossil sponges from about 580 million years ago in the Ediacaran period have been found in the Doushantuo Formation. These fossils, which include spicules, pinacocytes, porocytes, archeocytes, sclerocytes and the internal cavity, have been classified as demosponges. Fossils of glass sponges have been found from around 540 million years ago in rocks in Australia, China and Mongolia. Early Cambrian sponges from Mexico belonging to the genus *Kiwetinokia* show evidence of fusion of several smaller spicules to form a single large spicule. Calcium carbonate spicules of calcareous sponges have been found in Early Cambrian rocks from about 530 to 523 million years ago in Australia. Other probable demosponges have been found in the Early Cambrian Chengjiang fauna, from 525 to 520 million years ago. Freshwater sponges appear to be much younger, as the earliest known fossils date from the Mid-Eocene period about 48 to 40 million years ago. Although about 90% of modern sponges are demosponges, fossilized remains of this type are less common than those of other types because their skeletons are composed of relatively soft spongin that does not fossilize well. Earliest sponge symbionts are known from the early Silurian.

Archaeocyathids, which some classify as a type of coralline sponge, are very common fossils in rocks from the Early Cambrian about 530 to 520 million years ago, but apparently died out by the end of the Cambrian 490 million years ago. It has been suggested that they were produced by: sponges; cnidarians; algae; foraminiferans; a completely separate phylum of animals, Archaeocyatha; or even a completely separate kingdom of life, labeled Archaeata or Inferibionta. Since the 1990s archaeocyathids have been regarded as a distinctive group of sponges.

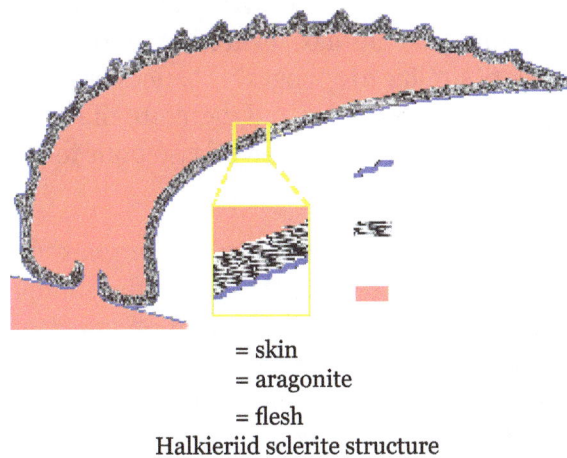

= skin
= aragonite
= flesh
Halkieriid sclerite structure

It is difficult to fit chancelloriids into classifications of sponges or more complex animals. An analysis in 1996 concluded that they were closely related to sponges on the grounds that the detailed structure of chancellorid sclerites ("armor plates") is similar to that of fibers of spongin, a collagen protein, in modern keratose (horny) demosponges such as *Darwinella*. However another analysis in 2002 concluded that chancelloriids are not sponges and may be intermediate between sponges and more complex animals, among other reasons because their skins were thicker and more tightly connected than those of sponges. In 2008 a detailed analysis of chancelloriids' sclerites concluded that they were very similar to those of halkieriids, mobile bilaterian animals that looked like slugs in chain mail and whose fossils are found in rocks from the very Early Cambrian to the Mid Cambrian. If this is correct, it would create a dilemma, as it is extremely unlikely that totally unrelated organisms could have developed such similar sclerites independently, but the huge difference in the structures of their bodies makes it hard to see how they could be closely related.

Relationships to other Animal Groups

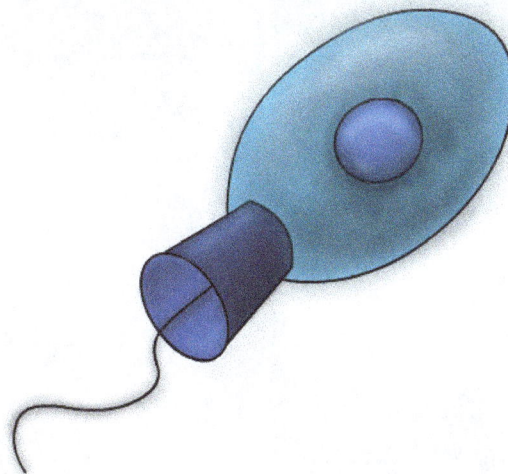

A choanoflagellate

In the 1990s sponges were widely regarded as a monophyletic group, all of them having descended from a common ancestor that was itself a sponge, and as the "sister-group" to all other metazoans (multi-celled animals), which themselves form a monophyletic group. On the other hand, some 1990s analyses also revived the idea that animals' nearest evolutionary relatives are choanoflagellates, single-celled organisms very similar to sponges' choanocytes – which would imply that most Metazoa evolved from very sponge-like ancestors and therefore that sponges may not be monophyletic, as the same sponge-like ancestors may have given rise both to modern sponges and to non-sponge members of Metazoa.

Analyses since 2001 have concluded that Eumetazoa (more complex than sponges) are more closely related to particular groups of sponges than to the rest of the sponges. Such conclusions imply that sponges are not monophyletic, because the last common ancestor of all sponges would also be a direct ancestor of the Eumetazoa, which are not sponges. A study in 2001 based on comparisons of ribosome DNA concluded that the most fundamental division within sponges was between glass sponges and the rest, and that Eumetazoa are more closely related to calcareous sponges, those with calcium carbonate spicules, than to other types of sponge. In 2007 one analysis based on comparisons of RNA and another based mainly on comparison of spicules concluded that demosponges and glass sponges are more closely related to each other than either is to calcareous sponges, which in turn are more closely related to Eumetazoa.

Other anatomical and biochemical evidence links the Eumetazoa with Homoscleromorpha, a subgroup of demosponges. A comparison in 2007 of nuclear DNA, excluding glass sponges and comb jellies, concluded that: Homoscleromorpha are most closely related to Eumetazoa; calcareous sponges are the next closest; the other demosponges are evolutionary "aunts" of these groups; and the chancelloriids, bag-like animals whose fossils are found in Cambrian rocks, may be sponges. The sperm of Homoscleromorpha share with those of Eumetazoa features that those of other sponges lack. In both Homoscleromorpha and Eumetazoa layers of cells are bound together by attachment to a carpet-like basal membrane composed mainly of "type IV" collagen, a form of collagen not found in other sponges – although the spongin fibers that reinforce the mesohyl of all demosponges is similar to "type IV" collagen.

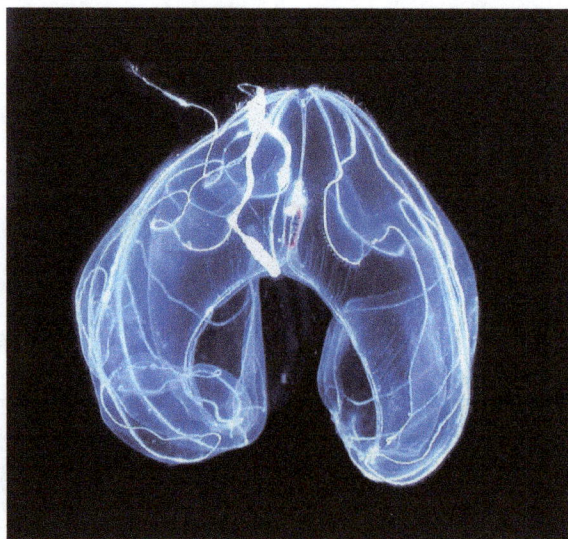

A comb jelly

The analyses described above concluded that sponges are closest to the ancestors of all Metazoa, of all multi-celled animals including both sponges and more complex groups. However, another comparison in 2008 of 150 genes in each of 21 genera, ranging from fungi to humans but including only two species of sponge, suggested that comb jellies (ctenophora) are the most basal lineage of the Metazoa included in the sample. If this is correct, either modern comb jellies developed their complex structures independently of other Metazoa, or sponges' ancestors were more complex and all known sponges are drastically simplified forms. The study recommended further analyses using a wider range of sponges and other simple Metazoa such as Placozoa. The results of such an analysis, published in 2009, suggest that a return to the previous view may be warranted. 'Family trees' constructed using a combination of all available data – morphological, developmental and molecular – concluded that the sponges are in fact a monophyletic group, and with the cnidarians form the sister group to the bilaterians.

A very large and internally consistent alignment of 1,719 proteins at the metazoan scale showed that (i) sponges – represented by Homoscleromorpha, Calcarea, Hexactinellida, and Demospongiae – are monophyletic, (ii) sponges are sister-group to all other multicellular animals, (iii) ctenophores emerge as the second-earliest branching animal lineage, and (iv) placozoans emerge as the third animal lineage, followed by cnidarians sister-group to bilaterians.

Use

By Dolphins

A report in 1997 described use of sponges as a tool by bottlenose dolphins in Shark Bay in Western Australia. A dolphin will attach a marine sponge to its rostrum, which is presumably then used to protect it when searching for food in the sandy sea bottom. The behavior, known as *sponging*, has only been observed in this bay, and is almost exclusively shown by females. A study in 2005 concluded that mothers teach the behavior to their daughters, and that all the sponge-users are closely related, suggesting that it is a fairly recent innovation.

By Humans

Natural sponges in Tarpon Springs, Florida

Display of natural sponges for sale on Kalymnos in Greece

Skeleton

The calcium carbonate or silica spicules of most sponge genera make them too rough for most uses, but two genera, *Hippospongia* and *Spongia*, have soft, entirely fibrous skeletons. Early Europeans used soft sponges for many purposes, including padding for helmets, portable drinking utensils and municipal water filters. Until the invention of synthetic sponges, they were used as cleaning tools, applicators for paints and ceramic glazes and discreet contraceptives. However, by the mid-20th century, over-fishing brought both the animals and the industry close to extinction.

Many objects with sponge-like textures are now made of substances not derived from poriferans. Synthetic sponges include personal and household cleaning tools, breast implants, and contraceptive sponges. Typical materials used are cellulose foam, polyurethane foam, and less frequently, silicone foam.

The luffa "sponge", also spelled *loofah*, which is commonly sold for use in the kitchen or the shower, is not derived from an animal but mainly from the fibrous "skeleton" of the sponge gourd (*Luffa aegyptiaca*, Cucurbitaceae).

Antibiotic Compounds

Sponges have medicinal potential due to the presence in sponges themselves or their microbial symbionts of chemicals that may be used to control viruses, bacteria, tumors and fungi.

Other Biologically Active Compounds

Lacking any protective shell or means of escape, sponges have evolved to synthesize a variety of unusual compounds. One such class is the oxidized fatty acid derivatives called oxylipins. Members of this family have been found to have anti-cancer, anti-bacterial and anti-fungal properties. One example isolated from the Okinawan *plakortis* sponges, plakoridine A, has shown potential as a cytotoxin to murine lymphoma cells.

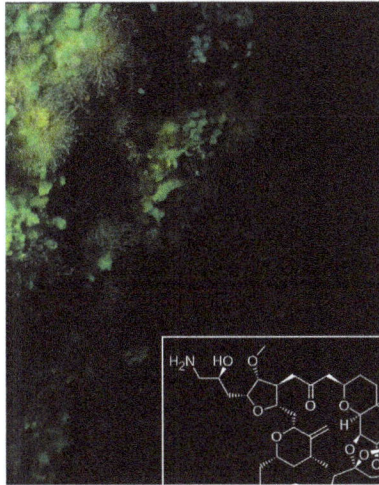

Halichondria produces the eribulin precursor halichondrin B

Placozoa

The Placozoa are a basal form of invertebrate. They are the simplest in structure of all non-parasitic multicellular animals (Animalia). They are generally classified as a single species, *Trichoplax adhaerens*, although there is enough genetic diversity that it is likely that there are multiple, morphologically similar species. Although they were first discovered in 1883 by the German zoologist, Franz Eilhard Schulze (1840-1921) and since the 1970s more systematically analyzed by the German protozoologist, Karl Gottlieb Grell (1912-1994), a common name does not yet exist for the taxon; the scientific name literally means "flat animals".

Biology

Trichoplax is a small, flattened, animal around 1 mm (0.039 in) across. Like an *Amoeba*, it has no regular outline, although the lower surface is somewhat concave, and the upper surface is always flattened. The body consists of an outer layer of simple epithelium enclosing a loose sheet of stellate cells resembling the mesenchyme of some more complex animals. The epithelial cells bear flagella, which the animal uses to help it creep along the seafloor.

The lower surface engulfs small particles of organic detritus, on which the animal feeds. It reproduces asexually, budding off smaller individuals, and the lower surface may also bud off eggs into the mesenchyme.

Evolutionary Relationships

There is no convincing fossil record of the placozoa, although the Ediacaran biota (Precambrian, 550 million years ago) organism *Dickinsonia* may be allied with this phylum.

Traditionally, classification was based on their level of organization: i.e. they possess no tissues or organs. However this may be as a result of secondary loss, so is inadequate to demark a clade. More recent work has attempted to classify them based on the DNA sequences in their genome; this has placed the phylum between the sponges and the eumetazoa. In such a feature-poor phylum, molecular data are considered to provide the most reliable approximation of the placozoans' phylogeny.

Functional-morphology Hypothesis

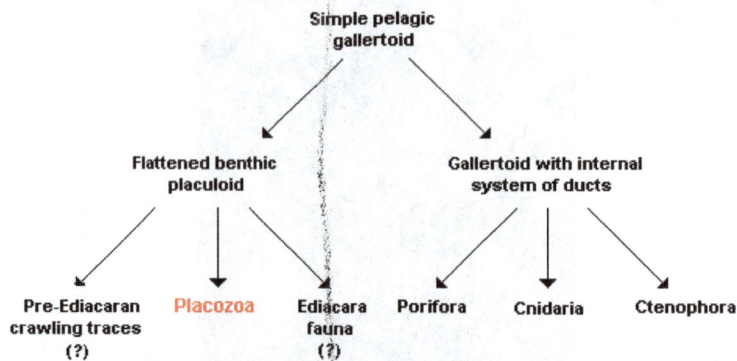

The Placozoa descending side by side with the sponges, cnidarians and ctenophores
from a gallertoid by processes of differentiation

On the basis of their simple structure, the Placozoa were frequently viewed as a model organism for the transition from unicellular organisms to the multicellular animals (Metazoa) and are thus considered a sister taxon to all other metazoans.

According to a functional-morphology model, all or most animals are descended from a *gallertoid*, a free-living (pelagic) sphere in seawater, consisting of a single ciliated layer of cells supported by a thin, noncellular separating layer, the basal lamina. The interior of the sphere is filled with contractile fibrous cells and a gelatinous extracellular matrix. Both the modern Placozoa and all other animals then descended from this multicellular beginning stage via two different processes:

- Infolding of the epithelium led to the formation of an internal system of ducts and thus to the development of a modified gallertoid from which the sponges (Porifera), Cnidaria and Ctenophora subsequently developed.

- Other gallertoids, according to this model, made the transition over time to a benthic mode of life; that is, their habitat has shifted from the open ocean to the floor (benthic zone). While the probability of encountering food, potential sexual partners, or predators is the same in all directions for animals floating freely in the water, there is a clear difference on the seafloor between the sides facing toward and away from the substrate, and between their orientation and the vertical direction perpendicular to the substrate. This results naturally in a selective advantage for flattening of the body, as of course can be seen in many benthic species. In the proposed functional-morphology model, the Placozoa, and possibly also several organisms known only from the fossil state, are descended from such a life form, which is now termed *placuloid*. Three different life strategies have accordingly led to three different lines of development:

  o Animals that live interstitially in the sand of the ocean floor were responsible for the fossil crawling traces that are considered the earliest evidence of animals and are detectable even prior to the dawn of the Ediacaran Period in geology. These are usually attributed to bilaterally symmetrical worms, but the hypothesis presented here views animals derived from placuloids, and thus close relatives of *Trichoplax adhaerens*, to be the producers of the traces.

- o Animals that incorporated algae as photosynthetically active endosymbionts, i.e. primarily obtaining their nutrients from their partners in symbiosis, were accordingly responsible for the mysterious creatures of the Ediacara fauna that are not assigned to any modern animal taxon and lived during the Ediacaran Period, before the start of the Paleozoic. Recent work has shown that some of the Ediacaran assemblages (e.g. Mistaken Point) were in deep water, below the photic zone, and that the organisms were not dependent on endosymbiotic photosynthesisers.

- o Animals that grazed on algal mats were ultimately the direct ancestors of the Placozoa. The advantages of an amoeboid multiplicity of shapes thus allowed a previously present basal lamina and a gelatinous extracellular matrix to be lost *secondarily*. Pronounced differentiation between the ventral surface facing the substrate and the dorsal, facing away from it, accordingly led to the physiologically distinct cell layers of *Trichoplax adhaerens* that can still be seen today. Consequently, these are analogous, but not homologous, to ectoderm and endoderm, the "external" and "internal" cell layers in eumetazoans; i.e. the structures corresponding functionally to one another have, according to the proposed hypothesis, no common evolutionary origin.

Should the analysis presented above turn out to be correct, *Trichoplax adhaerens* would be the oldest branch of the multicellular animals and a relic of the Ediacara fauna, or even the pre-Ediacara fauna. Due to the absence of extracellular matrix and basal lamina, the development potential of these animals, very successful in their ecological niche, was of course limited, which would explain the low rate of evolution, referred to as *bradytely*, of their phenotype, their outward form as adults.

This hypothesis was supported by a recent analysis of the *Trichoplax adhaerens* mitochondrial genome in comparison to those of other animals, The hypothesis was, however, rejected in a statistical analysis of the *Trichoplax adhaerens* whole genome sequence in comparison to the whole genome sequences of six other animals and two related non-animal species, but only at the $p=0.07$ level, which indicates a marginal level of statistical significance.

Epitheliozoa Hypothesis

A concept based on purely morphological characteristics pictures the Placozoa as the nearest relative of the animals with true tissues (Eumetazoa). The taxon they share, called the Epitheliozoa, is itself construed to be a sister group to the sponges (Porifera).

The principal support for such a relationship comes from special cell/cell junctions, the belt desmosomes, that occur not just in the Placozoa but in all animals except the sponges; they enable the cells to join together in an unbroken layer like the epitheloid of the Placozoa. *Trichoplax adhaerens* also shares the ventral gland cells with most eumetazoans. Both characteristics can be considered apomorphies, i.e. evolutionarily derived features, and thus form the basis of a common taxon for all animals that possess them.

One possible scenario inspired by the proposed hypothesis starts with the idea that the monociliated cells of the epitheloid in *Trichoplax adhaerens* evolved by reduction of the collars in the collar cells (choanocytes) of sponges as the ancestors of the Placozoa abandoned a filtering mode of life. The epitheloid would then have served as the precursor to the true epithelial tissue of the eumetazoans.

In contrast to the model based on functional morphology described earlier, in the Epitheliozoa concept the ventral and dorsal cell layers of the Placozoa are homologs of endoderm and ectoderm, the two basic embryonic cell layers of the eumetazoans — the digestive *gastrodermis* in the Cnidaria or the gut epithelium in the bilaterally symmetrical Bilateria may have developed from endoderm, whereas ectoderm is, among other things, the precursor to the external skin layer (epidermis). The interior space pervaded by a fiber syncytium in the Placozoa would then correspond to connective tissue in the other animals. It is uncertain whether the calcium ions stored in the syncytium are related to the lime skeletons of many cnidarians.

As noted above, this hypothesis was supported in a statistical analysis of the *Trichoplax adhaerens* whole genome sequence in comparison to the whole genome sequences of six other animals and two related non-animal species.

Eumetazoa Hypothesis

A third hypothesis, based primarily on molecular genetics, views the Placozoa as highly simplified eumetazoans. According to this, *Trichoplax adhaerens* is descended from considerably more complex animals that already had muscles and nerve tissues. Both tissue types, as well as the basal lamina of the epithelium, were accordingly lost more recently by radical secondary simplification.

Various studies in this regard so far yield differing results for identifying the exact sister group: in one case the Placozoa would qualify as the nearest relatives of the Cnidaria, while in another they would be a sister group to the Ctenophora, and occasionally they are placed directly next to the Bilateria.

An argument raised against the proposed scenario is that it leaves morphological features of the animals completely out of consideration. The extreme degree of simplification that would have to be postulated for the Placozoa in this model, moreover, is known only for parasitic organisms but would be difficult to explain functionally in a free-living species like *Trichoplax adhaerens*.

All versions of this hypothesis were rejected with high confidence in a statistical analysis of the *Trichoplax adhaerens* whole genome sequence in comparison to the whole genome sequences of six other animals and two related non-animal species.

Another DNA study suggests that these organisms are related to Cnidaria.

Eumetazoa

Eumetazoa is a clade comprising all major animal groups except sponges, placozoa, and several other extinct or obscure life forms, such as *Iotuba* and *Thectardis*. Characteristics of eumetazoans include true tissues organized into germ layers, the presence of neurons, and an embryo that goes through a gastrula stage. The clade is usually held to contain at least Ctenophora, Cnidaria, and Bilateria. Whether mesozoans belong is in dispute. Ctenophora now appear basal eumetazoa, and placozoa also appear to have emerged in eumetazoa. Eumetazoa would then be a basal Metazoan clade as sister of Porifera.

Some phylogenists have speculated the sponges and eumetazoans evolved separately from single-celled organisms, which would mean that the animal kingdom does not form a clade (a complete grouping of all organisms descended from a common ancestor). However, genetic studies and some morphological characteristics, like the common presence of choanocytes, support a common origin.

Eumetazoans are a major group of animals in the *Five Kingdoms* classification of Lynn Margulis and K. V. Schwartz, comprising the Radiata and Bilateria — all animals except the sponges, placozoans and mesozoans. When treated as a formal taxon Eumetazoa is typically ranked as a subkingdom. The name Metazoa has also been used to refer to this group, but more often refers to the Animalia as a whole. Many classification schemes do not include a subkingdom Eumetazoa.

Taxonomy

Over the last decade, the work of developmental biologists and molecular phylogeneticists spawned new ideas about bilaterian relationships resulting in a paradigm shift.

The current widely accepted hypothesis, based on molecular data (mostly 18S rRNA sequences), divides Bilateria into the following four superphylums: Deuterostomia, Ecdysozoa, Lophotrochozoa, and Platyzoa (sometimes included in Lophotrochozoa). The last three groups are also collectively known as Protostomia.

However, many skeptics emphasize the pitfalls and inconsistencies associated with the new data. Claus Nielsen, a professor of evolutionary invertebrate embryology at the Zoological Museum University of Copenhagen champions one of the most prominent alternative views based on morphological evidence. In his 2001 book *Animal Evolution: Interrelationships of the Living Phyla*, he maintains the traditional divisions of Protostomia and Deuterostomia.

Evolutionary Origins

It has been suggested that one type of molecular clock and one approach to interpretation of the fossil record both place the evolutionary origins of eumetazoa in the Ediacaran. However, the earliest eumetazoans may not have left a clear impact on the fossil record and other interpretations of molecular clocks suggest the possibility of an earlier origin. The discoverers of *Vernanimalcula* describe it as the fossil of a bilateral triploblastic animal that appeared at the end of the Marinoan glaciation prior to the Ediacaran Period, implying an even earlier origin for eumetazoans.

Radiata

Radiata is a taxonomic rank that has been used to classify radially symmetric animals. The term Radiata has united several different groupings of animals, some of which do not form a monophyletic group under current views of animal phylogeny. Because of this and problems of homoplasy associated with using body symmetry as a phylogenetic character, the term is used mostly in a historical context.

In the early 19th century, Georges Cuvier united ctenophores and cnidarians in the Radiata. Thomas Cavalier-Smith, in 1983, redefined Radiata as a subkingdom consisting of Myxozoa, Placozoa, Cnidaria and Ctenophora. Lynn Margulis and K. V. Schwartz later redefined Radiata in their *Five Kingdom* classification, this time including only Cnidaria and Ctenophora. This definition is similar to the historical descriptor coelenterata which has also been proposed as a group encompassing

Cnidaria and Ctenophora.

Although radial symmetry is usually given as a defining characteristic in animals that have been classified in this group, there are clear exceptions and qualifications. Echinoderms, for example, exhibit unmistaken bilateral symmetry as larvae. Ctenophores exhibit biradial or rotational symmetry, defined by tentacular and pharyngeal axes, on which two anal canals are located in two diametrically opposed quadrants. Some species within the cnidarian class Anthozoa are bilaterally symmetric (For example, *Nematostella vectensis*). It has been suggested that bilateral symmetry may have evolved before the split between Cnidaria and Bilateria, and that the radially symmetrical cnidarians have secondarily evolved radial symmetry, meaning the bilaterality in cnidarian species like *N. vectensis* has a primary origin.

Bilateria

The Bilateria or bilaterians, or triploblasts, are animals with bilateral symmetry, i.e., they have a head ("anterior") and a tail ("posterior") as well as a back ("dorsal") and a belly ("ventral"); therefore they also have a left side and a right side. In contrast, radially symmetrical animals like jellyfish have a topside and a downside, but no identifiable front or back.

The bilateria are a major group of animals, including the majority of phyla but not sponges, cnidarians, placozoans and ctenophores. For the most part, bilateral embryos are triploblastic, having three germ layers: endoderm, mesoderm, and ectoderm. Nearly all are bilaterally symmetrical, or approximately so; the most notable exception is the echinoderms, which achieve near-radial symmetry as adults, but are bilaterally symmetrical as larvae.

Except for a few phyla (i.e. flatworms and gnathostomulids), bilaterians have complete digestive tracts with a separate mouth and anus. Some bilaterians lack body cavities (acoelomates, i.e. Platyhelminthes, Gastrotricha and Gnathostomulida), while others display primary body cavities (deriving from the blastocoel, as pseudocoel) or secondary cavities (that appear *de novo*, for example the coelom).

Evolution

The hypothetical most recent common ancestor of all bilateria is termed the "Urbilaterian". The nature of the first bilaterian is a matter of debate. One side suggests that acoelomates gave rise to the other groups (planuloid-aceloid hypothesis by Graff, Metchnikoff, Hyman, or Salvini Plawen), while the other poses that the first bilaterian was a coelomate organism and the main acoelomate phyla (flatworms and gastrotrichs) have lost body cavities secondarily (the Archicoelomata hypothesis and its variations such as the Gastrea by Haeckel or Sedgwick, the Bilaterosgastrea by Gösta Jägersten (sv.), or the Trochaea by Nielsen).

The first evidence of bilateria in the fossil record comes from trace fossils in Ediacaran sediments, and the first *bona fide* bilaterian fossil is *Kimberella*, dating to 555 million years ago. Earlier fossils are controversial; the fossil *Vernanimalcula* may be the earliest known bilaterian, but may also represent an infilled bubble. Fossil embryos are known from around the time of *Vernanimalcula* (580 million years ago), but none of these have bilaterian affinities. Burrows believed to have been created by bilaterian life forms have been found in the Tacuarí Formation of Uruguay, and are believed to be at least 585 million years old.

Illustration of the different types of symmetry in lifeforms (Field Museum, Chicago). Bilateral forms can have heads. Lifeforms with other types of symmetry have corresponding organs, if not a head.

Phylogeny

There are two main superphyla (main lineages) of Bilateria. The deuterostomes include the echinoderms, hemichordates, chordates, and a few smaller phyla. The protostomes include most of the rest, such as arthropods, annelids, mollusks, flatworms, and so forth. There are a number of differences, most notably in how the embryo develops. In particular, the first opening of the embryo becomes the mouth in protostomes, and the anus in deuterostomes. Many taxonomists now recognize at least two more superphyla among the protostomes, Ecdysozoa (molting animals) and Lophotrochozoa (also referred to as Spiralia). Within the latter, some researchers also recognize another superphylum, Platyzoa, while others reject the Platyzoa monophyly. The arrow worms (Chaetognatha) have proven particularly difficult to classify, with some taxonomists placing them among the deuterostomes and others placing them among the protostomes. The two most recent studies to address the question of chaetognath origins support protostome affinities.

A modern (2011) consensus phylogeny for Bilateria is shown below, although the position of certain clades are still controversial and the tree has changed considerably between 2000 and 2010. Nodes marked with have received broad consensus. A prominent alternative tree is championed by Nielsen (2001).

Mesozoa

The Mesozoa (singular: mesozoon) are enigmatic, minuscule, worm-like parasites of marine invertebrates. As of 2012 it was still unclear whether they are degenerate platyhelminthes (flatworms) or truly-primitive, basal metazoans. Generally, these tiny, elusive creatures consist of a somatoderm (outer layer) of ciliated cells surrounding one or more reproductive cells. Decades ago, Mesozoa were classified as a phylum. Molecular phylogeny studies, however, have shown that the mysterious mesozoans are polyphyletic. That is, they consist of at least two unrelated groups.

As a result of these recent findings in molecular biology, the label mesozoan is now often applied informally, rather than as a formal taxon. Some workers previously classified Mesozoa as the sole phylum of the lonely subkingdom Agnotozoa. Cavalier-Smith argued that at least some of the mesozoans are in fact protistans, not animals.

In the 19th century, the Mesozoa were a wastebasket taxon for multicellular organisms which lacked the invaginating gastrula which was thought to define the Metazoa.

Evolution

Mesozoa were once thought to be evolutionary intermediate forms between Protozoans and Metazoans, but now they are thought to be degenerate or simplified metazoa. Their ciliated larva are similar to the miracidium of trematodes, and their internal multiplication is similar to what happens in the sprocysts of trematodes. Mesozoan DNA has a low GC-content (40%). This amount is similar to ciliates, but ciliates tend to be binucleate. Others relate mesozoa to a group including annelids, planarians, and nemerteans.

Groupings

The two main mesozoan groups are the Dicyemida and the Orthonectida. Other groups sometimes included in the Mesozoa are the Placozoa and the Monoblastozoa.

Monoblastozoans consist of a single description written in the 19th century of a species that has not been seen since. As such, many workers doubt that they are a real group. As described, the animal had only a single layer of tissue.

Rhombozoan Mesozoans

Rhombozoa, or dicyemid mesozoans, are found in the nephrid tracts of squid and octopuses. They range from a few millimeters long with twenty to thirty cells that include anterior attachment cells and a long central reproductive cell called an axial cell. This axial cell may develop asexually into vermiform juveniles or it may produce eggs and sperm that self-fertilize to produce a ciliated infusiform larva.

There are three genera: *Dicyema*, *Pseudicyema* and *Dicyemennea*.

Molecular evidence suggests that this phylum are derived from the *Lophotrochozoa*.

Orthonectid Mesozoans

Orthonectida are found in the body spaces of various marine invertebrates including tissue spaces, gonads, genitorespiratory bursae. This pathogen causes host castration of different species. Their actual phylogenetic position is uncertain, suggestions include as sister group to bilateral animals; as a relative of platyhelminthes and roundworms; as parasitic cnidarians similar to myxozoans and Polypodium hydriforme; and even as members of kingdom Protista, outside the animal kingdom.

The best known of Orthonectida is the parasite of brittle stars. The multinucleate syncytial stage lives within tissues and spaces of the gonad but can spread into arms. It causes the destruction of starfish ovary and eggs to cause castration (the male gonads are usually unaffected). The stages of the plasmodium develop into more plasmodia by simple fragmentation; at some point, they decide to go sexual. The syncytia are dioecious (either male or female), but young syncytia can fuse to produce both male and female. The males are ciliated and smaller than the females. The females and the males leave the starfish and mate in the sea. Tailed sperm enters the female and fertilizes the numerous oocytes. Each oocyst produces a small ciliated larva which makes its way to another star.

The genome of one of these species – *Intoshia linei* – has been sequenced.

Nephrozoa

Nephrozoa is a major clade of bilaterians including deuterostomes and protostomes. It is the extant sister clade of Xenacoelomorpha.

The majority of bilaterian animals are split into two groups, the protostomes and deuterostomes. Chordates (which include all the vertebrates) are deuterostomes. It seems very likely that the 555 million year old *Kimberella* was a member of the protostomes. If so, this means that the protostome and deuterostome lineages must have split some time before *Kimberella* appeared — at least 558 million years ago, and hence well before the start of the Cambrian 541 million years ago.

References

- Nielsen, C. (2002). Animal Evolution: Interrelationships of the Living Phyla (2nd ed.). England: Oxford University Press. ISBN 0-19-850682-1

- Lankester, R. (1877). Notes on the Embryology and classification of the Animal kingdom: comprising a revision of speculations relative to the origin and significance of the germ-layers. Quartely Journal of Microscopical Science (N.S.), No. 68: 399–454

- Hinde, R. T., (1998). "The Cnidaria and Ctenophora". In Anderson, D.T. Invertebrate Zoology. Oxford University Press. pp. 28–57. ISBN 0-19-551368-1. CS1 maint: Multiple names: authors list (link)

- "4 new species of 'killer' sponges discovered off Pacific coast". CBC News. April 19, 2014. Archived from the original on April 19, 2014. Retrieved 2014-09-04

- Leys, S. P. (2003). "The significance of syncytial tissues for the position of the Hexactinellida in the Metazoa". Integrative and Comparative Biology. 43 (1): 19–27. PMID 21680406. doi:10.1093/icb/43.1.19

- Eitel, Michael; Osigus, Hans-Jürgen; DeSalle, Rob; Schierwater, Bernd (2 April 2013). "Global Diversity of the Placozoa". PLoS ONE. 8 (4): e57131. doi:10.1371/journal.pone.0057131

- C. Hickman, C .P. (Jr.), Roberts, L. S., and Larson, A. (2001). Integrated Principles of Zoology (11 ed.). New York: McGraw-Hill. p. 247. ISBN 978-0-07-290961-6. CS1 maint: Multiple names: authors list (link)

- Vacelet, J. (2008). "A new genus of carnivorous sponges (Porifera: Poecilosclerida, Cladorhizidae) from the deep N-E Pacific, and remarks on the genus Neocladia" (PDF). Zootaxa. 1752: 57–65. Retrieved 2008-10-31

- Wulff, J. L (June 2008). "Collaboration among sponge species increases sponge diversity and abundance in a seagrass meadow". Marine Ecology. 29 (2): 193–204. doi:10.1111/j.1439-0485.2008.00224.x

- Takeuchi, Shinji; Ishibashi, Masami; Kobayashi, Junichi (1994). "Plakoridine A, a new tyramine-containing pyrrolidine alkaloid from the Okinawan marine sponge Plakortis sp". Journal of Organic Chemistry. 59 (13): 3712–3713. doi:10.1021/jo00092a039

- Smith, D. G. & Pennak, R. W. (2001). Pennak's Freshwater Invertebrates of the United States: Porifera to Crustacea (4 ed.). John Wiley and Sons. pp. 47–50. ISBN 0-471-35837-1

- Cavalcanti, F. F.; Klautau, M. (2011). "Solenoid: a new aquiferous system to Porifera". Zoomorphology. 130: 255–260. doi:10.1007/s00435-011-0139-7

- Nickel, M. (December 2004). "Kinetics and rhythm of body contractions in the sponge Tethya wilhelma (Porifera: Demospongiae)". Journal of Experimental Biology. 207 (Pt 26): 4515–4524. PMID 15579547. doi:10.1242/jeb.01289

- Imhoff, J. F. & Stöhr, R. (2003). "Sponge-Associated Bacteria". In Müller, W. E. G. Sponges (Porifera): Porifera. Springer. pp. 43–44. ISBN 3-540-00968-X

- "Demospongia". University of California Museum of Paleontology. Archived from the original on October 18, 2013. Retrieved 2008-11-27

- Etchells, L; Sardarian A.; Whitehead R. C. (18 April 2005). "A synthetic approach to the plakoridines modeled on a biogenetic theory". Tetrahedron Letters. 46 (16): 2803–2807. doi:10.1016/j.tetlet.2005.02.124

- Watling, L. (2007). "Predation on copepods by an Alaskan cladorhizid sponge". Journal of the Marine Biological Association of the United Kingdom. 87 (6): 1721–1726. doi:10.1017/S0025315407058560

- Margulis, Lynn (1988). Five Kingdoms: An illustrated Guide to the Phyla of Life on Earth. New York, NY, USA: W. H. FREEMAN AND COMPANY. ISBN 0716730278

- "Sponges". Cervical Barrier Advancement Society. 2004. Archived from the original on January 14, 2009. Retrieved 2006-09-17

- Dunn, Casey W.; Leys, Sally P.; Haddock, Steven H.D. (May 2015). "The hidden biology of sponges and ctenophores". Trends in Ecology & Evolution. 30 (5): 282–291. doi:10.1016/j.tree.2015.03.003

Vertebrate and Invertebrate Animals

3

Vertebrates are animals which have a spinal cord whereas animals without a spine are referred to as invertebrates. Some of the vertebrate animals are fish and amphibian. The topics discussed in the chapter are of great importance to broaden the existing knowledge on vertebrate and invertebrate animals.

Vertebrate

Vertebrates comprise all species of animals within the subphylum Vertebrata (chordates with backbones). Vertebrates represent the overwhelming majority of the phylum Chordata, with currently about 66,000 species described. Vertebrates include the jawless fish and the jawed vertebrates, which include the cartilaginous fish (sharks, rays, and ratfish) and the bony fish.

A bony fish clade known as the lobe-finned fishes is included with tetrapods, which are further divided into amphibians, reptiles, mammals, and birds. Extant vertebrates range in size from the frog species *Paedophryne amauensis*, at as little as 7.7 mm (0.30 in), to the blue whale, at up to 33 m (108 ft). Vertebrates make up less than five percent of all described animal species; the rest are invertebrates, which lack vertebral columns.

The vertebrates traditionally include the hagfish, which do not have proper vertebrae due to their loss in evolution, though their closest living relatives, the lampreys, do. Hagfish do, however, possess a cranium. For this reason, the vertebrate subphylum is sometimes referred to as "Craniata" when discussing morphology.

Molecular analysis since 1992 has suggested that hagfish are most closely related to lampreys, and so also are vertebrates in a monophyletic sense. Others consider them a sister group of vertebrates in the common taxon of craniata.

Etymology

The word origin of *vertebrate* derives from the Latin word *vertebratus* (Pliny), meaning *joint of the spine*. The Proto-Indo-European language origins are still unclear.

Vertebrate is derived from the word *vertebra*, which refers to any of the bones or segments of the spinal column.

Anatomy and Morphology

All vertebrates are built along the basic chordate body plan: a stiff rod running through the length of the animal (vertebral column and/or notochord), with a hollow tube of nervous tissue (the spinal cord) above it and the gastrointestinal tract below.

In all vertebrates, the mouth is found at, or right below, the anterior end of the animal, while the anus opens to the exterior before the end of the body. The remaining part of the body continuing after the anus forms a tail with vertebrae and spinal cord, but no gut.

Vertebral Column

The defining characteristic of a vertebrate is the vertebral column, in which the notochord (a stiff rod of uniform composition) found in all chordates has been replaced by a segmented series of stiffer elements (vertebrae) separated by mobile joints (intervertebral discs, derived embryonically and evolutionarily from the notochord).

However, a few vertebrates have secondarily lost this anatomy, retaining the notochord into adulthood, such as the sturgeon and coelacanth. Jawed vertebrates are typified by paired appendages (fins or legs, which may be secondarily lost), but this trait is not required in order for an animal to be a vertebrate.

Fossilized skeleton of *Diplodocus carnegii*, showing an extreme example of the backbone that characterizes the vertebrates.

Gills

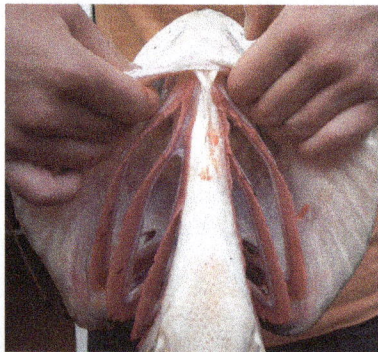

Gill arches bearing gills in a pike

All basal vertebrates breathe with gills. The gills are carried right behind the head, bordering the posterior margins of a series of openings from the pharynx to the exterior. Each gill is supported by a cartilagenous or bony gill arch. The bony fish have three pairs of arches, cartilaginous fish have five to seven pairs, while the primitive jawless fish have seven. The vertebrate ancestor no doubt had more arches than this, as some of their chordate relatives have more than 50 pairs of gills.

In amphibians and some primitive bony fishes, the larvae bear external gills, branching off from the gill arches. These are reduced in adulthood, their function taken over by the gills proper in fishes and by lungs in most amphibians. Some amphibians retain the external larval gills in adulthood, the complex internal gill system as seen in fish apparently being irrevocably lost very early in the evolution of tetrapods.

While the higher vertebrates lack gills, the gill arches form during fetal development, and form the basis of essential structures such as jaws, the thyroid gland, the larynx, the *columella* (corresponding to the stapes in mammals) and, in mammals, the malleus and incus.

Central Nervous System

The central nervous system of vertebrates is based on a hollow nerve cord running along the length of the animal. Of particular importance and unique to vertebrates is the presence of neural crest cells. These are progenitors of stem cells, and critical to coordinating the functions of cellular components. Neural crest cells migrate through the body from the nerve cord during development, and initiate the formation of neural ganglia and structures such as the jaws and skull.

The vertebrates are the only chordate group to exhibit cephalisation, the concentration of brain functions in the head. A slight swelling of the anterior end of the nerve cord is found in the lancelet, a chordate, though it lacks the eyes and other complex sense organs comparable to those of vertebrates. Other chordates do not show any trends towards cephalisation.

A peripheral nervous system branches out from the nerve cord to innervate the various systems. The front end of the nerve tube is expanded by a thickening of the walls and expansion of the central canal of spinal cord into three primary brain vesicles: The prosencephalon (forebrain), mesencephalon (midbrain) and rhombencephalon (hindbrain), further differentiated in the various vertebrate groups. Two laterally placed eyes form around outgrowths from the midbrain, except in hagfish, though this may be a secondary loss. The forebrain is well developed and subdivided in most tetrapods, while the midbrain dominate in many fish and some salamanders. Vesicles of the forebrain are usually paired, giving rise to hemispheres like the cerebral hemispheres in mammals.

The resulting anatomy of the central nervous system, with a single hollow nerve cord topped by a series of (often paired) vesicles, is unique to vertebrates. All invertebrates with well-developed brains, such as insects, spiders and squids, have a ventral rather than dorsal system of ganglions, with a split brain stem running on each side of the mouth or gut.

Evolutionary History

First Vertebrates

The early vertebrate *Haikouichthys*

Vertebrates originated about 525 million years ago during the Cambrian explosion, which saw the rise in organism diversity. The earliest known vertebrate is believed to be the *Myllokunmingia*.

Another early vertebrate is *Haikouichthys ercaicunensis*. Unlike the other fauna that dominated the Cambrian, these groups had the basic vertebrate body plan: a notochord, rudimentary vertebrae, and a well-defined head and tail. All of these early vertebrates lacked jaws in the common sense and relied on filter feeding close to the seabed. A vertebrate group of uncertain phylogeny, small-eel-like conodonts, are known from microfossils of their paired tooth segments from the late Cambrian to the end of the Triassic.

From Fish to Amphibians

Acanthostega, a fish-like early labyrinthodont.

The first jawed vertebrates appeared in the latest Ordovician and became common in the Devonian, often known as the "Age of Fishes". The two groups of bony fishes, the actinopterygii and sarcopterygii, evolved and became common. The Devonian also saw the demise of virtually all jawless fishes, save for lampreys and hagfish, as well as the Placodermi, a group of armoured fish that dominated the entirety of that period since the late Silurian. The Devonian also saw the rise of the first labyrinthodonts, which was a transitional form between fishes and amphibians.

Mesozoic Vertebrates

Amniotes branched from labyrinthodonts in the subsequent Carboniferous period. The Parareptilia and synapsid amniotes were common during the late Paleozoic, while diapsids became dominant during the Mesozoic. In the sea, the bony fishes became dominant. The birds are a derived form of dinosaurs evolved in the Jurassic. The demise of the non-avian dinosaurs at the end of the Cretaceous allowed for the expansion of mammals, which had evolved from the therapsids, a group of synapsid amniotes, during the late Triassic Period.

After the Mesozoic

The Cenozoic world has seen great diversification of bony fishes, frogs, birds and mammals.

Over half of all living vertebrate species (about 32,000 species) are fish (non-tetrapod craniates), a diverse set of lineages that inhabit all the world's aquatic ecosystems, from snow minnows (Cypriniformes) in Himalayan lakes at elevations over 4,600 metres (15,100 feet) to flatfishes (order Pleuronectiformes) in the Challenger Deep, the deepest ocean trench at about 11,000 metres (36,000 feet). Fishes of myriad varieties are the main predators in most of the world's water bodies, both freshwater and marine. The rest of the vertebrate species are tetrapods, a single lineage that includes amphibians (with roughly 7,000 species); mammals (with approximately 5,500 species); and reptiles and birds (with about 20,000 species divided evenly between the two classes). Tetrapods comprise the dominant megafauna of most terrestrial environments and also include many partially or fully aquatic groups (e.g., sea snakes, penguins, cetaceans).

Classification

There are several ways of classifying animals. Evolutionary systematics relies on anatomy, physiology and evolutionary history, which is determined through similarities in anatomy and, if possible, the genetics of organisms. Phylogenetic classification is based solely on phylogeny. Evolutionary systematics gives an overview; phylogenetic systematics gives detail. The two systems are thus complementary rather than opposed.

Traditional Classification

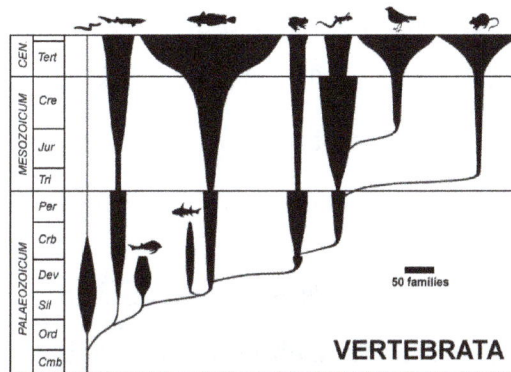

Traditional spindle diagram of the evolution of the vertebrates at class level

Conventional classification has living vertebrates grouped into seven classes based on traditional interpretations of gross anatomical and physiological traits. This classification is the one most commonly encountered in school textbooks, overviews, non-specialist, and popular works. The extant vertebrates are:

- Subphylum Vertebrata

  o Class Agnatha (jawless fishes)

  o Class Chondrichthyes (cartilaginous fishes)

  o Class Osteichthyes (bony fishes)

  o Class Amphibia (amphibians)

  o Class Reptilia (reptiles)

  o Class Aves (birds)

  o Class Mammalia (mammals)

In addition to these, there are two classes of extinct armoured fishes, the Placodermi and the Acanthodii.

Other ways of classifying the vertebrates have been devised, particularly with emphasis on the phylogeny of early amphibians and reptiles. An example based on Janvier (1981, 1997), Shu *et al.* (2003), and Benton (2004) is given here:

- Subphylum Vertebrata

- o *Palaeospondylus*

- o Superclass Agnatha or Cephalaspidomorphi (lampreys and other jawless fishes)

- o Infraphylum Gnathostomata (vertebrates with jaws)

    - ◇ Class †Placodermi (extinct armoured fishes)

    - ◇ Class Chondrichthyes (cartilaginous fishes)

    - ◇ Class †Acanthodii (extinct spiny "sharks")

    - ◇ Superclass Osteichthyes (bony vertebrates)

        - * Class Actinopterygii (ray-finned bony fishes)

        - * Class Sarcopterygii (lobe-finned fishes, tetrapods are inside this class)

        - * Class Amphibia (amphibians, some ancestral to the amniotes)- now a paraphyletic group

        - * Class Synapsida (mammals are placed inside this thought to be extinct taxon)

        - * Class Sauropsida (reptiles, birds are inside this group in a monophyletic way)

While this traditional classification is orderly, most of the groups are paraphyletic, i.e. do not contain all descendants of the class's common ancestor. For instance, descendants of the first reptiles include modern reptiles, as well as mammals and birds. Most of the classes listed are not "complete" (and are therefore paraphyletic) taxa, meaning they do not include all the descendants of the first representative of the group. For example, the agnathans have given rise to the jawed vertebrates; the bony fishes have given rise to the land vertebrates; the traditional "amphibians" have given rise to the reptiles (traditionally including the synapsids, or mammal-like "reptiles"), which in turn have given rise to the mammals and birds. Most scientists working with vertebrates use a classification based purely on phylogeny, organized by their known evolutionary history and sometimes disregarding the conventional interpretations of their anatomy and physiology.

Phylogenetic Relationships

In phylogenetic taxonomy, the relationships between animals are not typically divided into ranks, but illustrated as a nested "family tree" known as a cladogram. Phylogenetic groups are given definitions based on their relationship to one another, rather than purely on physical traits, such as the presence of a backbone. This nesting pattern is often combined with traditional taxonomy (as above), in a practice known as evolutionary taxonomy.

Reproductive Systems

Nearly all vertebrates undergo sexual reproduction. They produce haploid gametes by meiosis. The smaller, motile gametes are spermatozoa and the larger, non-motile gametes are ova. These fuse by the process of fertilisation to form diploid zygotes, which develop into new individuals.

Inbreeding

During sexual reproduction, mating with a close relative (inbreeding) often leads to inbreeding depression. Inbreeding depression is considered to be largely due to expression of deleterious recessive mutations. The effects of inbreeding have been studied in many vertebrate species.

In several species of fish, inbreeding was found to decrease reproductive success.

Inbreeding was observed to increase juvenile mortality in 11 small animal species.

A common breeding practice for pet dogs is mating between close relatives (e.g. between half- and full siblings). This practice generally has a negative effect on measures of reproductive success, including decreased litter size and puppy survival.

Incestuous matings in birds result in severe fitness costs due to inbreeding depression (e.g. reduction in hatchability of eggs and reduced progeny survival).

Inbreeding Avoidance

As a result of the negative fitness consequences of inbreeding, vertebrate species have evolved mechanisms to avoid inbreeding. Numerous inbreeding avoidance mechanisms operating prior to mating have been described.

Toads and many other amphibians display breeding site fidelity. Individuals that return to natal ponds to breed will likely encounter siblings as potential mates. Although incest is possible, *Bufo americanus* siblings rarely mate. These toads likely recognize and actively avoid close kins as mates. Advertisement vocalizations by males appear to serve as cues by which females recognize their kin.

Inbreeding avoidance mechanisms can also operate subsequent to copulation. In guppies, a post-copulatory mechanism of inbreeding avoidance occurs based on competition between sperm of rival males for achieving fertilization. In competitions between sperm from an unrelated male and from a full sibling male, a significant bias in paternity towards the unrelated male was observed.

When female sand lizards mate with two or more males, sperm competition within the female's reproductive tract may occur. Active selection of sperm by females appears to occur in a manner that enhances female fitness. On the basis of this selective process, the sperm of males that are more distantly related to the female are preferentially used for fertilization, rather than the sperm of close relatives. This preference may enhance the fitness of progeny by reducing inbreeding depression.

Outcrossing

Mating with unrelated or distantly related members of the same species is generally thought to provide the advantage of masking deleterious recessive mutations in progeny. Vertebrates have evolved numerous diverse mechanisms for avoiding close inbreeding and promoting outcrossing.

Outcrossing as a way of avoiding inbreeding depression, has been especially well studied in birds. For instance, inbreeding depression occurs in the great tit when the offspring are produced as a result of a mating between close relatives. In natural populations of the great tit (*Parus major*), inbreeding is avoided by dispersal of individuals from their birthplace, which reduces the chance of mating with a close relative.

The purple-crowned fairywren females paired with related males may undertake extra-pair matings that can reduce the negative effects of inbreeding. However, there are ecological and demographic constraints on extra pair matings. Nevertheless, 46% of broods produced by incestuously paired females contained extra-pair young.

Southern pied babblers (*Turdoides bicolor*) appear to avoid inbreeding in two ways. The first is through dispersal, and the second is by avoiding familiar group members as mates. Although both males and females disperse locally, they move outside the range where genetically related individuals are likely to be encountered. Within their group, individuals only acquire breeding positions when the opposite-sex breeder is unrelated.

Cooperative breeding in birds typically occurs when offspring, usually males, delay dispersal from their natal group in order to remain with the family to help rear younger kin. Female offspring rarely stay at home, dispersing over distances that allow them to breed independently, or to join unrelated groups.

Parthenogenesis

Parthenogenesis is a natural form of reproduction in which growth and development of embryos occur without fertilization.

Reproduction in squamate reptiles is ordinarily sexual, with males having a ZZ pair of sex determining chromosomes, and females a ZW pair. However, various species, including the Colombian Rainbow boa (*Epicrates maurus*), *Agkistrodon contortrix* (copperhead snake) and *Agkistrodon piscivorus* (cotton mouth snake) can also reproduce by facultative parthenogenesis -that is, they are capable of switching from a sexual mode of reproduction to an asexual mode- resulting in production of WW female progeny. The WW females are likely produced by terminal automixis.

Mole salamanders are an ancient (2.4-3.8 million year-old) unisexual vertebrate lineage. In the polyploid unisexual mole salamander females, a premeiotic endomitotic event doubles the number of chromosomes. As a result, the mature eggs produced subsequent to the two meiotic divisions have the same ploidy as the somatic cells of the female salamander. Synapsis and recombination during meiotic prophase I in these unisexual females is thought to ordinarily occur between identical sister chromosomes and occasionally between homologous chromosomes. Thus little, if any, genetic variation is produced. Recombination between homeologous chromosomes occurs only rarely, if at all. Since production of genetic variation is weak, at best, it is unlikely to provide a benefit sufficient to account for the long-term maintenance of meiosis in these organisms. However, meiosis may have been maintained during evolution by the efficient recombinational repair of DNA damages that meiosis provides, an advantage that could be realized at each generation.

Self-fertilization

The mangrove killifish (*Kryptolebias marmoratus*) produces both eggs and sperm by meiosis and routinely reproduces by self-fertilisation. The capacity for selfing in these fishes has apparently persisted for at least several hundred thousand years. Each individual hermaphrodite normally fertilizes itself when an egg and sperm that it has produced by an internal organ unite inside the fish's body. In nature, this mode of reproduction can yield highly homozygous lines composed of individuals so genetically uniform as to be, in effect, identical to one another. Although inbreeding, especially in the extreme form of self-fertilization, is ordinarily regarded as detrimental because it leads to expression of deleterious recessive alleles, self-fertilization does provide the benefit of *fertilization assurance* (reproductive assurance) at each generation.

Various Vertebrate Animals

Fish

A fish is any member of a group of animals that consist of all gill-bearing aquatic craniate animals that lack limbs with digits. They form a sister group to the tunicates, together forming the olfactores. Included in this definition are the living hagfish, lampreys, and cartilaginous and bony fish as well as various extinct related groups. Tetrapods emerged within lobe-finned fishes, so cladistically they are fish as well. However, traditionally fish are rendered obsolete or paraphyletic by excluding the tetrapods (i.e., the amphibians, reptiles, birds and mammals which all descended from within the same ancestry). Because in this manner the term "fish" is defined negatively as a paraphyletic group, it is not considered a formal taxonomic grouping in systematic biology. The traditional term pisces (also ichthyes) is considered a typological, but not a phylogenetic classification.

The earliest organisms that can be classified as fish were soft-bodied chordates that first appeared during the Cambrian period. Although they lacked a true spine, they possessed notochords which allowed them to be more agile than their invertebrate counterparts. Fish would continue to evolve through the Paleozoic era, diversifying into a wide variety of forms. Many fish of the Paleozoic developed external armor that protected them from predators. The first fish with jaws appeared in the Silurian period, after which many (such as sharks) became formidable marine predators rather than just the prey of arthropods.

Most fish are ectothermic ("cold-blooded"), allowing their body temperatures to vary as ambient temperatures change, though some of the large active swimmers like white shark and tuna can hold a higher core temperature. Fish are abundant in most bodies of water. They can be found in nearly all aquatic environments, from high mountain streams (e.g., char and gudgeon) to the abyssal and even hadal depths of the deepest oceans (e.g., gulpers and anglerfish). With 33,100 described species, fish exhibit greater species diversity than any other group of vertebrates.

Fish are an important resource for humans worldwide, especially as food. Commercial and subsistence fishers hunt fish in wild fisheries or farm them in ponds or in cages in the ocean. They are also caught by recreational fishers, kept as pets, raised by fishkeepers, and exhibited in public aquaria. Fish have had a role in culture through the ages, serving as deities, religious symbols, and as the subjects of art, books and movies.

Evolution

Fish do not represent a monophyletic group, and therefore the "evolution of fish" is not studied as a single event.

Early fish from the fossil record are represented by a group of small, jawless, armored fish known as ostracoderms. Jawless fish lineages are mostly extinct. An extant clade, the lampreys may approximate ancient pre-jawed fish. The first jaws are found in Placodermi fossils. The diversity of jawed vertebrates may indicate the evolutionary advantage of a jawed mouth. It is unclear if the advantage of a hinged jaw is greater biting force, improved respiration, or a combination of factors.

Dunkleosteus was a gigantic, 10-metre (33 ft) long prehistoric fish of class Placodermi.

Fish may have evolved from a creature similar to a coral-like sea squirt, whose larvae resemble primitive fish in important ways. The first ancestors of fish may have kept the larval form into adulthood (as some sea squirts do today), although perhaps the reverse is the case.

Taxonomy

Fish are a paraphyletic group: that is, any clade containing all fish also contains the tetrapods, which are not fish. For this reason, groups such as the "Class Pisces" seen in older reference works are no longer used in formal classifications.

Traditional classification divide fish into three extant classes, and with extinct forms sometimes classified within the tree, sometimes as their own classes:

- Class Agnatha (jawless fish)
  - Subclass Cyclostomata (hagfish and lampreys)
  - Subclass Ostracodermi (armoured jawless fish) †
- Class Chondrichthyes (cartilaginous fish)
  - Subclass Elasmobranchii (sharks and rays)
  - Subclass Holocephali (chimaeras and extinct relatives)
- Class Placodermi (armoured fish) †
- Class Acanthodii ("spiny sharks", sometimes classified under bony fishes)†

Leedsichthys (left), of the subclass Actinopterygii, is the largest known fish, with estimates in 2005 putting its maximum size at 16 metres (52 ft)

- Class Osteichthyes (bony fish)
  - Subclass Actinopterygii (ray finned fishes)
  - Subclass Sarcopterygii (fleshy finned fishes, ancestors of tetrapods)

The above scheme is the one most commonly encountered in non-specialist and general works. Many of the above groups are paraphyletic, in that they have given rise to successive groups: Agnathans are ancestral to Chondrichthyes, who again have given rise to Acanthodiians, the ancestors of Osteichthyes. With the arrival of phylogenetic nomenclature, the fishes has been split up into a more detailed scheme, with the following major groups:

- Class Myxini (hagfish)
- Class Pteraspidomorphi † (early jawless fish)
- Class Thelodonti †
- Class Anaspida †
- Class Petromyzontida or Hyperoartia
  - Petromyzontidae (lampreys)
- Class Conodonta (conodonts) †
- Class Cephalaspidomorphi † (early jawless fish)
  - (unranked) Galeaspida †
  - (unranked) Pituriaspida †
  - (unranked) Osteostraci †
- Infraphylum Gnathostomata (jawed vertebrates)
  - Class Placodermi † (armoured fish)
  - Class Chondrichthyes (cartilaginous fish)
  - Class Acanthodii † (spiny sharks)
  - Superclass Osteichthyes (bony fish)
    - Class Actinopterygii (ray-finned fish)

- Subclass Chondrostei
  - Order Acipenseriformes (sturgeons and paddlefishes)
  - Order Polypteriformes (reedfishes and bichirs).
- Subclass Neopterygii
  - Infraclass Holostei (gars and bowfins)
  - Infraclass Teleostei (many orders of common fish)
- Class Sarcopterygii (lobe-finned fish)
  - Subclass Actinistia (coelacanths)
  - Subclass Dipnoi (lungfish)

† – indicates extinct taxon

Some palaeontologists contend that because Conodonta are chordates, they are primitive fish. For a fuller treatment of this taxonomy.

The position of hagfish in the phylum Chordata is not settled. Phylogenetic research in 1998 and 1999 supported the idea that the hagfish and the lampreys form a natural group, the Cyclostomata, that is a sister group of the Gnathostomata.

The various fish groups account for more than half of vertebrate species. There are almost 28,000 known extant species, of which almost 27,000 are bony fish, with 970 sharks, rays, and chimeras and about 108 hagfish and lampreys. A third of these species fall within the nine largest families; from largest to smallest, these families are Cyprinidae, Gobiidae, Cichlidae, Characidae, Loricariidae, Balitoridae, Serranidae, Labridae, and Scorpaenidae. About 64 families are monotypic, containing only one species. The final total of extant species may grow to exceed 32,500.

Diversity

Examples of the Major Classes of Fish

Agnatha (Pacific hagfish)

The term "fish" most precisely describes any non-tetrapod craniate (i.e. an animal with a skull and in most cases a backbone) that has gills throughout life and whose limbs, if any, are in the shape of fins. Unlike groupings such as birds or mammals, fish are not a single clade but a paraphyletic collection of taxa, including hagfishes, lampreys, sharks and rays, ray-finned fish, coelacanths, and lungfish. Indeed, lungfish and coelacanths are closer relatives of tetrapods (such as mammals, birds, amphibians, etc.) than of other fish such as ray-finned fish or sharks, so the last common ancestor of all fish is also an ancestor to tetrapods. As paraphyletic groups are no longer recognised in modern systematic biology, the use of the term "fish" as a biological group must be avoided.

Chondrichthyes (Horn shark)

Many types of aquatic animals commonly referred to as "fish" are not fish in the sense given above; examples include shellfish, cuttlefish, starfish, crayfish and jellyfish. In earlier times, even biologists did not make a distinction – sixteenth century natural historians classified also seals, whales, amphibians, crocodiles, even hippopotamuses, as well as a host of aquatic invertebrates, as fish. However, according to the definition above, all mammals, including cetaceans like whales and dolphins, are not fish. In some contexts, especially in aquaculture, the true fish are referred to as finfish (or fin fish) to distinguish them from these other animals.

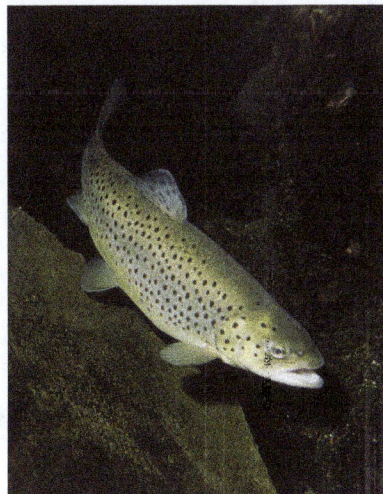

Actinopterygii (Brown trout)

A typical fish is ectothermic, has a streamlined body for rapid swimming, extracts oxygen from water using gills or uses an accessory breathing organ to breathe atmospheric oxygen, has two sets of paired fins, usually one or two (rarely three) dorsal fins, an anal fin, and a tail fin, has jaws, has skin that is usually covered with scales, and lays eggs.

Sarcopterygii (Coelacanth)

Each criterion has exceptions. Tuna, swordfish, and some species of sharks show some warm-blooded adaptations—they can heat their bodies significantly above ambient water temperature. Streamlining and swimming performance varies from fish such as tuna, salmon, and jacks that can cover 10–20 body-lengths per second to species such as eels and rays that swim no more than 0.5 body-lengths per second. Many groups of freshwater fish extract oxygen from the air as well as from the water using a variety of different structures. Lungfish have paired lungs similar to those of tetrapods, gouramis have a structure called the labyrinth organ that performs a similar function, while many catfish, such as *Corydoras* extract oxygen via the intestine or stomach. Body shape and the arrangement of the fins is highly variable, covering such seemingly un-fishlike forms as seahorses, pufferfish, anglerfish, and gulpers. Similarly, the surface of the skin may be naked (as in moray eels), or covered with scales of a variety of different types usually defined as placoid (typical of sharks and rays), cosmoid (fossil lungfish and coelacanths), ganoid (various fossil fish but also living gars and bichirs), cycloid, and ctenoid (these last two are found on most bony fish). There are even fish that live mostly on land or lay their eggs on land near water. Mudskippers feed and interact with one another on mudflats and go underwater to hide in their burrows. A single, an undescribed species of *Phreatobius*, has been called a true "land fish" as this worm-like catfish strictly lives among waterlogged leaf litter. Many species live in underground lakes, underground rivers or aquifers and are popularly known as cavefish.

Fish come in many shapes and sizes. This is a sea dragon, a close relative of the seahorse. Their leaf-like appendages enable them to blend in with floating seaweed.

Fish range in size from the huge 16-metre (52 ft) whale shark to the tiny 8-millimetre (0.3 in) stout infantfish.

Fish species diversity is roughly divided equally between marine (oceanic) and freshwater ecosystems. Coral reefs in the Indo-Pacific constitute the center of diversity for marine fishes, whereas continental freshwater fishes are most diverse in large river basins of tropical rainforests, especially the Amazon, Congo, and Mekong basins. More than 5,600 fish species inhabit Neotropical freshwaters alone, such that Neotropical fishes represent about 10% of all vertebrate species on the Earth. Exceptionally rich sites in the Amazon basin, such as Cantão State Park, can contain more freshwater fish species than occur in all of Europe.

Anatomy and Physiology

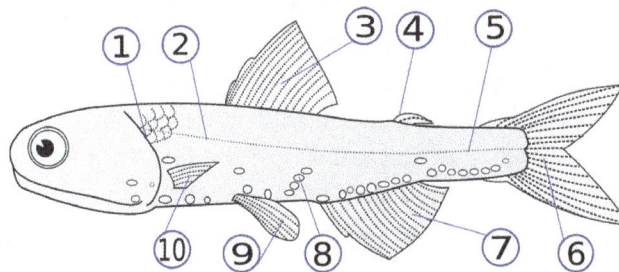

The anatomy of *Lampanyctodes hectoris*
(1) – operculum (gill cover), (2) – lateral line, (3) – dorsal fin, (4) – fat fin, (5) – caudal peduncle, (6) – caudal fin,
(7) – anal fin, (8) – photophores, (9) – pelvic fins (paired), (10) – pectoral fins (paired)

Respiration

Gills

Most fish exchange gases using gills on either side of the pharynx. Gills consist of threadlike structures called filaments. Each filament contains a capillary network that provides a large surface area for exchanging oxygen and carbon dioxide. Fish exchange gases by pulling oxygen-rich water through their mouths and pumping it over their gills. In some fish, capillary blood flows in the opposite direction to the water, causing countercurrent exchange. The gills push the oxygen-poor water out through openings in the sides of the pharynx. Some fish, like sharks and lampreys, possess multiple gill openings. However, bony fish have a single gill opening on each side. This opening is hidden beneath a protective bony cover called an operculum.

Juvenile bichirs have external gills, a very primitive feature that they share with larval amphibians.

Air Breathing

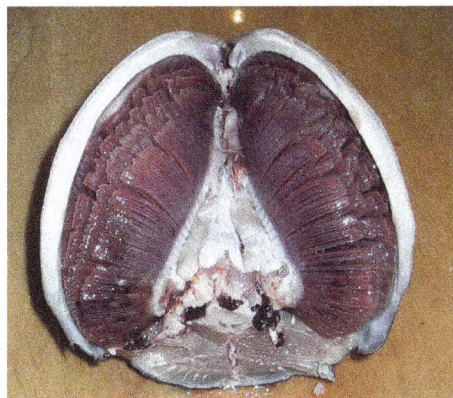

Tuna gills inside the head. The fish head is oriented snout-downwards, with the view looking towards the mouth.

Fish from multiple groups can live out of the water for extended periods. Amphibious fish such as the mudskipper can live and move about on land for up to several days, or live in stagnant or otherwise oxygen depleted water. Many such fish can breathe air via a variety of mechanisms. The skin of anguillid eels may absorb oxygen directly. The buccal cavity of the electric eel may breathe air. Catfish of the families Loricariidae, Callichthyidae, and Scoloplacidae absorb air through their digestive tracts. Lungfish, with the exception of the Australian lungfish, and bichirs have paired lungs similar to those of tetrapods and must surface to gulp fresh air through the mouth and pass spent air out through the gills. Gar and bowfin have a vascularized swim bladder that functions in the same way. Loaches, trahiras, and many catfish breathe by passing air through the gut. Mudskippers breathe by absorbing oxygen across the skin (similar to frogs). A number of fish have evolved so-called accessory breathing organs that extract oxygen from the air. Labyrinth fish (such as gouramis and bettas) have a labyrinth organ above the gills that performs this function. A few other fish have structures resembling labyrinth organs in form and function, most notably snakeheads, pikeheads, and the Clariidae catfish family.

Breathing air is primarily of use to fish that inhabit shallow, seasonally variable waters where the water's oxygen concentration may seasonally decline. Fish dependent solely on dissolved oxygen, such as perch and cichlids, quickly suffocate, while air-breathers survive for much longer, in some cases in water that is little more than wet mud. At the most extreme, some air-breathing fish are able to survive in damp burrows for weeks without water, entering a state of aestivation (summertime hibernation) until water returns.

Air breathing fish can be divided into obligate air breathers and facultative air breathers. Obligate air breathers, such as the African lungfish, *must* breathe air periodically or they suffocate. Facultative air breathers, such as the catfish *Hypostomus plecostomus*, only breathe air if they need to and will otherwise rely on their gills for oxygen. Most air breathing fish are facultative air breathers that avoid the energetic cost of rising to the surface and the fitness cost of exposure to surface predators.

Circulation

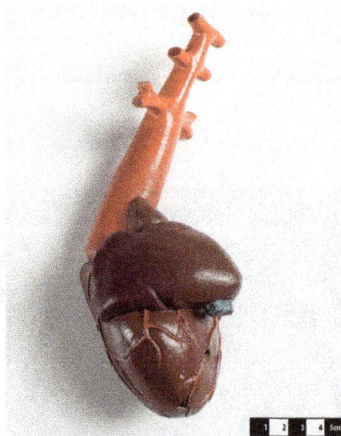

Didactic model of a fish heart.

Fish have a closed-loop circulatory system. The heart pumps the blood in a single loop throughout the body. In most fish, the heart consists of four parts, including two chambers and an entrance and exit. The first part is the sinus venosus, a thin-walled sac that collects blood from the fish's veins before allowing it to flow to the second part, the atrium, which is a large muscular chamber. The atrium serves as a one-way antechamber, sends blood to the third part, ventricle. The ventricle is another thick-walled, muscular chamber and it pumps the blood, first to the fourth part, bulbus arteriosus, a large tube, and then out of the heart. The bulbus arteriosus connects to the aorta, through which blood flows to the gills for oxygenation.

Digestion

Jaws allow fish to eat a wide variety of food, including plants and other organisms. Fish ingest food through the mouth and break it down in the esophagus. In the stomach, food is further digested and, in many fish, processed in finger-shaped pouches called pyloric caeca, which secrete digestive enzymes and absorb nutrients. Organs such as the liver and pancreas add enzymes and various chemicals as the food moves through the digestive tract. The intestine completes the process of digestion and nutrient absorption.

Excretion

As with many aquatic animals, most fish release their nitrogenous wastes as ammonia. Some of the wastes diffuse through the gills. Blood wastes are filtered by the kidneys.

Saltwater fish tend to lose water because of osmosis. Their kidneys return water to the body. The reverse happens in freshwater fish: they tend to gain water osmotically. Their kidneys produce dilute urine for excretion. Some fish have specially adapted kidneys that vary in function, allowing them to move from freshwater to saltwater.

Scales

The scales of fish originate from the mesoderm (skin); they may be similar in structure to teeth.

Sensory and Nervous System

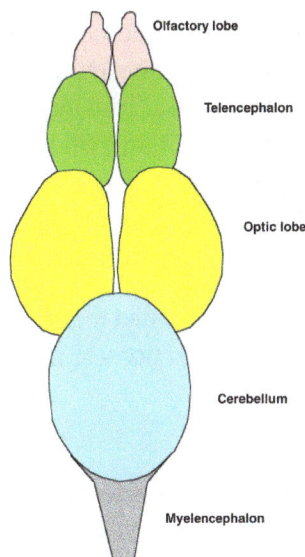

Dorsal view of the brain of the rainbow trout

Central Nervous System

Fish typically have quite small brains relative to body size compared with other vertebrates, typically one-fifteenth the brain mass of a similarly sized bird or mammal. However, some fish have relatively large brains, most notably mormyrids and sharks, which have brains about as massive relative to body weight as birds and marsupials.

Fish brains are divided into several regions. At the front are the olfactory lobes, a pair of structures that receive and process signals from the nostrils via the two olfactory nerves. The olfactory lobes are very large in fish that hunt primarily by smell, such as hagfish, sharks, and catfish. Behind the olfactory lobes is the two-lobed telencephalon, the structural equivalent to the cerebrum in higher vertebrates. In fish the telencephalon is concerned mostly with olfaction. Together these structures form the forebrain.

Connecting the forebrain to the midbrain is the diencephalon (in the diagram, this structure is below the optic lobes and consequently not visible). The diencephalon performs functions associated with hormones and homeostasis. The pineal body lies just above the diencephalon. This structure detects light, maintains circadian rhythms, and controls color changes.

The midbrain or mesencephalon contains the two optic lobes. These are very large in species that hunt by sight, such as rainbow trout and cichlids.

The hindbrain or metencephalon is particularly involved in swimming and balance. The cerebellum is a single-lobed structure that is typically the biggest part of the brain. Hagfish and lampreys have relatively small cerebellae, while the mormyrid cerebellum is massive and apparently involved in their electrical sense.

The brain stem or myelencephalon is the brain's posterior. As well as controlling some muscles and body organs, in bony fish at least, the brain stem governs respiration and osmoregulation.

Sense Organs

Most fish possess highly developed sense organs. Nearly all daylight fish have color vision that is at least as good as a human's. Many fish also have chemoreceptors that are responsible for extraordinary senses of taste and smell. Although they have ears, many fish may not hear very well. Most fish have sensitive receptors that form the lateral line system, which detects gentle currents and vibrations, and senses the motion of nearby fish and prey. Some fish, such as catfish and sharks, have the Ampullae of Lorenzini, organs that detect weak electric currents on the order of millivolt. Other fish, like the South American electric fishes Gymnotiformes, can produce weak electric currents, which they use in navigation and social communication.

Fish orient themselves using landmarks and may use mental maps based on multiple landmarks or symbols. Fish behavior in mazes reveals that they possess spatial memory and visual discrimination.

Vision

Vision is an important sensory system for most species of fish. Fish eyes are similar to those of terrestrial vertebrates like birds and mammals, but have a more spherical lens. Their retinas generally have both rods and cones (for scotopic and photopic vision), and most species have colour vision.

Some fish can see ultraviolet and some can see polarized light. Amongst jawless fish, the lamprey has well-developed eyes, while the hagfish has only primitive eyespots. Fish vision shows adaptation to their visual environment, for example deep sea fishes have eyes suited to the dark environment.

Hearing

Hearing is an important sensory system for most species of fish. Fish sense sound using their lateral lines and their ears.

Capacity for Pain

Experiments done by William Tavolga provide evidence that fish have pain and fear responses. For instance, in Tavolga's experiments, toadfish grunted when electrically shocked and over time they came to grunt at the mere sight of an electrode.

In 2003, Scottish scientists at the University of Edinburgh and the Roslin Institute concluded that rainbow trout exhibit behaviors often associated with pain in other animals. Bee venom and acetic acid injected into the lips resulted in fish rocking their bodies and rubbing their lips along the sides and floors of their tanks, which the researchers concluded were attempts to relieve pain, similar to what mammals would do. Neurons fired in a pattern resembling human neuronal patterns.

Professor James D. Rose of the University of Wyoming claimed the study was flawed since it did not provide proof that fish possess "conscious awareness, particularly a kind of awareness that is meaningfully like ours". Rose argues that since fish brains are so different from human brains, fish are probably not conscious in the manner humans are, so that reactions similar to human reactions to pain instead have other causes. Rose had published a study a year earlier arguing that fish cannot feel pain because their brains lack a neocortex. However, animal behaviorist Temple Grandin argues that fish could still have consciousness without a neocortex because "different species can use different brain structures and systems to handle the same functions."

Animal welfare advocates raise concerns about the possible suffering of fish caused by angling. Some countries, such as Germany have banned specific types of fishing, and the British RSPCA now formally prosecutes individuals who are cruel to fish.

Muscular System

Swim bladder of a rudd (*Scardinius erythrophthalmus*)

Most fish move by alternately contracting paired sets of muscles on either side of the backbone. These contractions form S-shaped curves that move down the body. As each curve reaches the back fin, backward force is applied to the water, and in conjunction with the fins, moves the fish forward. The fish's fins function like an airplane's flaps. Fins also increase the tail's surface area, increasing speed. The streamlined body of the fish decreases the amount of friction from the water. Since body tissue is denser than water, fish must compensate for the difference or they will sink. Many bony fish have an internal organ called a swim bladder that adjusts their buoyancy through manipulation of gases.

A great white shark off Isla Guadalupe

Homeothermy

Although most fish are exclusively ectothermic, there are exceptions.

Certain species of fish maintain elevated body temperatures. Endothermic teleosts (bony fish) are all in the suborder Scombroidei and include the billfishes, tunas, and one species of "primitive" mackerel (*Gasterochisma melampus*). All sharks in the family Lamnidae – shortfin mako, long fin mako, white, porbeagle, and salmon shark – are endothermic, and evidence suggests the trait exists in family Alopiidae (thresher sharks). The degree of endothermy varies from the billfish, which warm only their eyes and brain, to bluefin tuna and porbeagle sharks who maintain body temperatures elevated in excess of 20 °C above ambient water temperatures. Endothermy, though metabolically costly, is thought to provide advantages such as increased muscle strength, higher rates of central nervous system processing, and higher rates of digestion.

Reproductive System

Organs: 1. Liver, 2. Gas bladder, 3. Roe, 4. Pyloric caeca, 5. Stomach, 6. Intestine

Fish reproductive organs include testicles and ovaries. In most species, gonads are paired organs of similar size, which can be partially or totally fused. There may also be a range of secondary organs that increase reproductive fitness.

In terms of spermatogonia distribution, the structure of teleosts testes has two types: in the most common, spermatogonia occur all along the seminiferous tubules, while in atherinomorph fish they are confined to the distal portion of these structures. Fish can present cystic or semi-cystic spermatogenesis in relation to the release phase of germ cells in cysts to the seminiferous tubules lumen.

Fish ovaries may be of three types: gymnovarian, secondary gymnovarian or cystovarian. In the first type, the oocytes are released directly into the coelomic cavity and then enter the ostium, then through the oviduct and are eliminated. Secondary gymnovarian ovaries shed ova into the coelom from which they go directly into the oviduct. In the third type, the oocytes are conveyed to the exterior through the oviduct. Gymnovaries are the primitive condition found in lungfish, sturgeon, and bowfin. Cystovaries characterize most teleosts, where the ovary lumen has continuity with the oviduct. Secondary gymnovaries are found in salmonids and a few other teleosts.

Oogonia development in teleosts fish varies according to the group, and the determination of oogenesis dynamics allows the understanding of maturation and fertilization processes. Changes in the nucleus, ooplasm, and the surrounding layers characterize the oocyte maturation process.

Postovulatory follicles are structures formed after oocyte release; they do not have endocrine function, present a wide irregular lumen, and are rapidly reabsorbed in a process involving the apoptosis of follicular cells. A degenerative process called follicular atresia reabsorbs vitellogenic oocytes not spawned. This process can also occur, but less frequently, in oocytes in other development stages.

Some fish, like the California sheephead, are hermaphrodites, having both testes and ovaries either at different phases in their life cycle or, as in hamlets, have them simultaneously.

Over 97% of all known fish are oviparous, that is, the eggs develop outside the mother's body. Examples of oviparous fish include salmon, goldfish, cichlids, tuna, and eels. In the majority of these species, fertilisation takes place outside the mother's body, with the male and female fish shedding their gametes into the surrounding water. However, a few oviparous fish practice internal fertilization, with the male using some sort of intromittent organ to deliver sperm into the genital opening of the female, most notably the oviparous sharks, such as the horn shark, and oviparous rays, such as skates. In these cases, the male is equipped with a pair of modified pelvic fins known as claspers.

Marine fish can produce high numbers of eggs which are often released into the open water column. The eggs have an average diameter of 1 millimetre (0.039 in).

| Egg of lamprey | Egg of catshark (mermaids' purse) | Egg of bullhead shark | Egg of chimaera | Ovary of fish (Corumbatá) |

The newly hatched young of oviparous fish are called larvae. They are usually poorly formed, carry a large yolk sac (for nourishment), and are very different in appearance from juvenile and adult specimens. The larval period in oviparous fish is relatively short (usually only several weeks), and larvae rapidly grow and change appearance and structure (a process termed metamorphosis) to become juveniles. During this transition larvae must switch from their yolk sac to feeding on zooplankton prey, a process which depends on typically inadequate zooplankton density, starving many larvae.

In ovoviviparous fish the eggs develop inside the mother's body after internal fertilization but receive little or no nourishment directly from the mother, depending instead on the yolk. Each embryo develops in its own egg. Familiar examples of ovoviviparous fish include guppies, angel sharks, and coelacanths.

Some species of fish are viviparous. In such species the mother retains the eggs and nourishes the embryos. Typically, viviparous fish have a structure analogous to the placenta seen in mammals connecting the mother's blood supply with that of the embryo. Examples of viviparous fish include the surf-perches, splitfins, and lemon shark. Some viviparous fish exhibit oophagy, in which the developing embryos eat other eggs produced by the mother. This has been observed primarily among sharks, such as the shortfin mako and porbeagle, but is known for a few bony fish as well, such as the halfbeak *Nomorhamphus ebrardtii*. Intrauterine cannibalism is an even more unusual mode of vivipary, in which the largest embryos eat weaker and smaller siblings. This behavior is also most commonly found among sharks, such as the grey nurse shark, but has also been reported for *Nomorhamphus ebrardtii*.

Aquarists commonly refer to ovoviviparous and viviparous fish as livebearers.

Diseases

Like other animals, fish suffer from diseases and parasites. To prevent disease they have a variety of defenses. *Non-specific* defenses include the skin and scales, as well as the mucus layer secreted by the epidermis that traps and inhibits the growth of microorganisms. If pathogens breach these defenses, fish can develop an inflammatory response that increases blood flow to the infected region and delivers white blood cells that attempt to destroy pathogens. Specific defenses respond to particular pathogens recognised by the fish's body, i.e., an immune response. In recent years, vaccines have become widely used in aquaculture and also with ornamental fish, for example furunculosis vaccines in farmed salmon and koi herpes virus in koi.

Some species use cleaner fish to remove external parasites. The best known of these are the Bluestreak cleaner wrasses of the genus *Labroides* found on coral reefs in the Indian and Pacific oceans. These small fish maintain so-called "cleaning stations" where other fish congregate and perform specific movements to attract the attention of the cleaners. Cleaning behaviors have been observed in a number of fish groups, including an interesting case between two cichlids of the same genus, *Etroplus maculatus*, the cleaner, and the much larger *Etroplus suratensis*.

Immune System

Immune organs vary by type of fish. In the jawless fish (lampreys and hagfish), true lymphoid organs are absent. These fish rely on regions of lymphoid tissue within other organs to produce immune cells. For example, erythrocytes, macrophages and plasma cells are produced in the anterior kidney (or pronephros) and some areas of the gut (where granulocytes mature.) They resemble primitive bone marrow in hagfish. Cartilaginous fish (sharks and rays) have a more advanced immune system. They have three specialized organs that are unique to Chondrichthyes; the epigonal organs (lymphoid tissue similar to mammalian bone) that surround the gonads, the Leydig's organ within the walls of their esophagus, and a spiral valve in their intestine. These organs house typical immune cells (granulocytes, lymphocytes and plasma cells). They also possess an identifiable thymus and a well-developed spleen (their most important immune organ) where various lymphocytes, plasma cells and macrophages develop and are stored. Chondrostean fish (sturgeons, paddlefish, and bichirs) possess a major site for the production of granulocytes within a mass that is associated with the meninges (membranes surrounding the central nervous system.) Their heart is frequently covered with tissue that contains lymphocytes, reticular cells and a small number of macrophages. The chondrostean kidney is an important hemopoietic organ; where erythrocytes, granulocytes, lymphocytes and macrophages develop.

Like chondrostean fish, the major immune tissues of bony fish (or teleostei) include the kidney (especially the anterior kidney), which houses many different immune cells. In addition, teleost fish possess a thymus, spleen and scattered immune areas within mucosal tissues (e.g. in the skin, gills, gut and gonads). Much like the mammalian immune system, teleost erythrocytes, neutrophils and granulocytes are believed to reside in the spleen whereas lymphocytes are the major cell type found in the thymus. In 2006, a lymphatic system similar to that in mammals was described in one species of teleost fish, the zebrafish. Although not confirmed as yet, this system presumably will be where naive (unstimulated) T cells accumulate while waiting to encounter an antigen.

B and T lymphocytes bearing immunoglobulins and T cell receptors, respectively, are found in all jawed fishes. Indeed, the adaptive immune system as a whole evolved in an ancestor of all jawed vertebrate.

Conservation

The 2006 IUCN Red List names 1,173 fish species that are threatened with extinction. Included are species such as Atlantic cod, Devil's Hole pupfish, coelacanths, and great white sharks. Because fish live underwater they are more difficult to study than terrestrial animals and plants, and information about fish populations is often lacking. However, freshwater fish seem particularly threatened because they often live in relatively small water bodies. For example, the Devil's Hole pupfish occupies only a single 3 by 6 metres (10 by 20 ft) pool.

Overfishing

A Whale shark, the world's largest fish, is classified as Vulnerable.

Overfishing is a major threat to edible fish such as cod and tuna. Overfishing eventually causes population (known as stock) collapse because the survivors cannot produce enough young to replace those removed. Such commercial extinction does not mean that the species is extinct, merely that it can no longer sustain a fishery.

One well-studied example of fishery collapse is the Pacific sardine *Sadinops sagax caerulues* fishery off the California coast. From a 1937 peak of 790,000 long tons (800,000 t) the catch steadily declined to only 24,000 long tons (24,000 t) in 1968, after which the fishery was no longer economically viable.

The main tension between fisheries science and the fishing industry is that the two groups have different views on the resiliency of fisheries to intensive fishing. In places such as Scotland, New-foundland, and Alaska the fishing industry is a major employer, so governments are predisposed to support it. On the other hand, scientists and conservationists push for stringent protection, warning that many stocks could be wiped out within fifty years.

Habitat Destruction

A key stress on both freshwater and marine ecosystems is habitat degradation including water pollution, the building of dams, removal of water for use by humans, and the introduction of exotic species. An example of a fish that has become endangered because of habitat change is the pallid sturgeon, a North American freshwater fish that lives in rivers damaged by human activity.

Exotic Species

Introduction of non-native species has occurred in many habitats. One of the best studied examples is the introduction of Nile perch into Lake Victoria in the 1960s. Nile perch gradually exterminated the lake's 500 endemic cichlid species. Some of them survive now in captive breeding programmes, but others are probably extinct. Carp, snakeheads, tilapia, European perch, brown trout, rainbow trout, and sea lampreys are other examples of fish that have caused problems by being introduced into alien environments.

Importance to Humans

Economic Importance

These fish-farming ponds were created as a cooperative project in a rural village.

Throughout history, humans have utilized fish as a food source. Historically and today, most fish protein has come by means of catching wild fish. However, aquaculture, or fish farming, which has been practiced since about 3,500 BCE. in China, is becoming increasingly important in many nations. Overall, about one-sixth of the world's protein is estimated to be provided by fish. That proportion is considerably elevated in some developing nations and regions heavily dependent on the sea. In a similar manner, fish have been tied to trade.

Catching fish for the purpose of food or sport is known as fishing, while the organized effort by humans to catch fish is called a fishery. Fisheries are a huge global business and provide income for millions of people. The annual yield from all fisheries worldwide is about 154 million tons, with popular species including herring, cod, anchovy, tuna, flounder, and salmon. However, the term fishery is broadly applied, and includes more organisms than just fish, such as mollusks and crustaceans, which are often called "fish" when used as food.

Recreation

Fish have been recognized as a source of beauty for almost as long as used for food, appearing in cave art, being raised as ornamental fish in ponds, and displayed in aquariums in homes, offices, or public settings.

Recreational fishing is fishing for pleasure or competition; it can be contrasted with commercial fishing, which is fishing for profit. The most common form of recreational fishing is done with a rod, reel, line, hooks and any one of a wide range of baits. Angling is a method of fishing, specifically the practice of catching fish by means of an "angle" (hook). Anglers must select the right hook, cast accurately, and retrieve at the right speed while considering water and weather conditions, species, fish response, time of the day, and other factors.

Culture

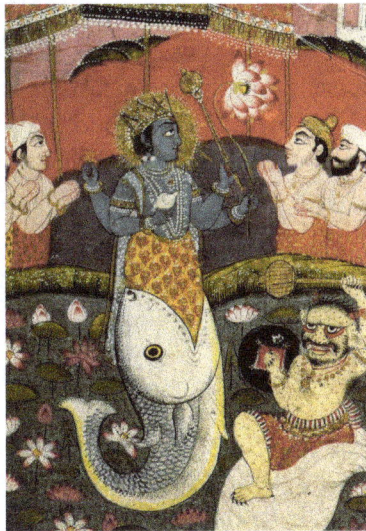
Avatar of Vishnu as a Matsya

Fish feature prominently in art and literature, in movies such as *Finding Nemo* and books such as *The Old Man and the Sea*. Large fish, particularly sharks, have frequently been the subject of horror movies and thrillers, most notably the novel *Jaws*, which spawned a series of films of the same name that in turn inspired similar films or parodies such as *Shark Tale* and *Snakehead Terror*. Piranhas are shown in a similar light to sharks in films such as *Piranha*; however, contrary to popular belief, the red-bellied piranha is actually a generally timid scavenger species that is unlikely to harm humans. In the Book of Jonah a "great fish" swallowed Jonah the Prophet. Legends of half-human, half-fish mermaids have featured in folklore, including the stories of Hans Christian Andersen.

Fish themes have symbolic significance in many religions. The fish is used often as a symbol by Christians to represent Jesus, or Christianity in general; the gospels also refer to "fishers of men" and feeding the multitude. In the dhamma of Buddhism the fish symbolize happiness as they have complete freedom of movement in the water. Often drawn in the form of carp which are regarded in the Orient as sacred on account of their elegant beauty, size and life-span. Among the deities said to take the form of a fish are Ika-Roa of the Polynesians, Dagon of various ancient Semitic peoples, the shark-gods of Hawaii and Matsya of the Hindus.

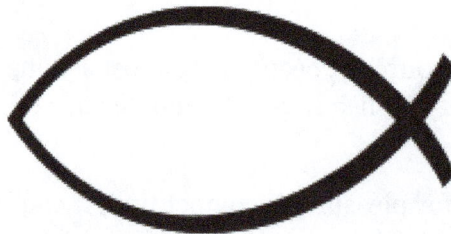

The ichthus is a Christian symbol of a fish signifying that the person who uses it is a Christian.

The astrological symbol Pisces is based on a constellation of the same name, but there is also a second fish constellation in the night sky, Piscis Austrinus.

Terminology

Fish or Fishes

Though often used interchangeably, in biology these words have different meanings. *Fish* is used as a singular noun, or as a plural to describe multiple individuals from a single species. *Fishes* is used to describe different species or species groups. Thus a pond that contained a single species might be said to contain 120 fish. But if the pond contained a total of 120 fish from three different species, it would be said to contain three fishes. The distinction is similar to that between people and peoples.

True Fish and Finfish

- In biology, the term *fish* is most strictly used to describe any animal with a backbone that has gills throughout life and has limbs, if any, in the shape of fins. Many types of aquatic animals with common names ending in "fish" are not fish in this sense; examples include shellfish, cuttlefish, starfish, crayfish and jellyfish. In earlier times, even biologists did not make a distinction—sixteenth century natural historians classified also seals, whales, amphibians, crocodiles, even hippopotamuses, as well as a host of aquatic invertebrates, as fish.

- In fisheries, the term *fish* is used as a collective term, and includes mollusks, crustaceans and any aquatic animal which is harvested.

- The strict biological definition of a fish, above, is sometimes called a *true fish*. True fish are also referred to as *finfish* or *fin fish* to distinguish them from other aquatic life harvested in fisheries or aquaculture.

Shoal or School

These goldband fusiliers are schooling because their swimming is synchronised

A random assemblage of fish merely using some localised resource such as food or nesting sites is known simply as an *aggregation*. When fish come together in an interactive, social grouping, then they may be forming either a *shoal* or a *school* depending on the degree of organisation. A *shoal* is a loosely organised group where each fish swims and forages independently but is attracted to other members of the group and adjusts its behaviour, such as swimming speed, so that it remains close to the other members of the group. *Schools* of fish are much more tightly organised, synchronising their swimming so that all fish move at the same speed and in the same direction. Shoaling and schooling behaviour is believed to provide a variety of advantages.

Examples:

- Cichlids congregating at lekking sites form an *aggregation*.
- Many minnows and characins form *shoals*.
- Anchovies, herrings and silversides are classic examples of *schooling* fish.

While the words "school" and "shoal" have different meanings within biology, the distinctions are often ignored by non-specialists who treat the words as synonyms. Thus speakers of British English commonly use "shoal" to describe any grouping of fish, and speakers of American English commonly use "school" just as loosely.

Amphibian

Amphibians are ectothermic, tetrapod vertebrates of the class Amphibia. Modern amphibians are all Lissamphibia. They inhabit a wide variety of habitats, with most species living within terrestrial, fossorial, arboreal or freshwater aquatic ecosystems. Thus amphibians typically start out as larvae living in water, but some species have developed behavioural adaptations to bypass this. The young generally undergo metamorphosis from larva with gills to an adult air-breathing form with lungs. Amphibians use their skin as a secondary respiratory surface and some small terrestrial salamanders and frogs lack lungs and rely entirely on their skin. They are superficially similar to lizards but, along with mammals and birds, reptiles are amniotes and do not require water bodies in which to breed. With their complex reproductive needs and permeable skins, amphibians are often ecological indicators; in recent decades there has been a dramatic decline in amphibian populations for many species around the globe.

The earliest amphibians evolved in the Devonian period from sarcopterygian fish with lungs and bony-limbed fins, features that were helpful in adapting to dry land. They diversified and became dominant during the Carboniferous and Permian periods, but were later displaced by reptiles and other vertebrates. Over time, amphibians shrank in size and decreased in diversity, leaving only the modern subclass Lissamphibia.

The three modern orders of amphibians are Anura (the frogs and toads), Urodela (the salamanders), and Apoda (the caecilians). The number of known amphibian species is approximately 7,000, of which nearly 90% are frogs. The smallest amphibian (and vertebrate) in the world is a frog from New Guinea (*Paedophryne amauensis*) with a length of just 7.7 mm (0.30 in). The largest living amphibian is the 1.8 m (5 ft 11 in) Chinese giant salamander (*Andrias davidianus*), but this is dwarfed by the extinct 9 m (30 ft) *Prionosuchus* from the middle Permian of Brazil. The study of amphibians is called batrachology, while the study of both reptiles and amphibians is called herpetology.

Classification

The word "amphibian" is derived from the Ancient Greek term (*amphíbios*), which means "both kinds of life" meaning "of both kinds" and meaning "life". The term was initially used as a general adjective for animals that could live on land or in water, including seals and otters. Traditionally, the class Amphibia includes all tetrapod vertebrates that are not amniotes. Amphibia in its widest sense (*sensu lato*) was divided into three subclasses, two of which are extinct:

The world's smallest known vertebrate, *Paedophryne amauensis*, sitting on a U.S. dime, which is 17.9 mm in diameter, for scale

- Subclass Lepospondyli[†] (small Paleozoic group, which may actually be more closely related to amniotes than Lissamphibia)

- Subclass Temnospondyli[†] (diverse Paleozoic and early Mesozoic grade)

- Subclass Lissamphibia (all modern amphibians, including frogs, toads, salamanders, newts and caecilians)
  - Salientia (frogs, toads and relatives): Jurassic to present—6,200 current species in 53 families

- Caudata (salamanders, newts and relatives): Jurassic to present—652 current species in 9 families

- Gymnophiona (caecilians and relatives): Jurassic to present—192 current species in 10 families

Triadobatrachus massinoti, a proto-frog from the Early Triassic of Madagascar

The actual number of species in each group depends on the taxonomic classification followed. The two most common systems are the classification adopted by the website AmphibiaWeb, University of California, Berkeley and the classification by herpetologist Darrel Frost and the American Museum of Natural History, available as the online reference database "Amphibian Species of the World". The numbers of species cited above follows Frost and the total number of known amphibian species is over 7,000, of which nearly 90% are frogs.

With the phylogenetic classification, the taxon Labyrinthodontia has been discarded as it is a polyparaphyletic group without unique defining features apart from shared primitive characteristics. Classification varies according to the preferred phylogeny of the author and whether they use a stem-based or a node-based classification. Traditionally, amphibians as a class are defined as all tetrapods with a larval stage, while the group that includes the common ancestors of all living amphibians (frogs, salamanders and caecilians) and all their descendants is called Lissamphibia. The phylogeny of Paleozoic amphibians is uncertain, and Lissamphibia may possibly fall within extinct groups, like the Temnospondyli (traditionally placed in the subclass Labyrinthodontia) or the Lepospondyli, and in some analyses even in the amniotes. This means that advocates of phylogenetic nomenclature have removed a large number of basal Devonian and Carboniferous amphibian-type tetrapod groups that were formerly placed in Amphibia in Linnaean taxonomy, and included them elsewhere under cladistic taxonomy. If the common ancestor of amphibians and amniotes is included in Amphibia, it becomes a paraphyletic group.

All modern amphibians are included in the subclass Lissamphibia, which is usually considered a clade, a group of species that have evolved from a common ancestor. The three modern orders are Anura (the frogs and toads), Caudata (or Urodela, the salamanders), and Gymnophiona (or Apoda, the caecilians). It has been suggested that salamanders arose separately from a Temnospondyl-like ancestor, and even that caecilians are the sister group of the advanced reptiliomorph amphibians, and thus of amniotes. Although the fossils of several older proto-frogs with primitive characteristics are known, the oldest "true frog" is *Prosalirus bitis*, from the Early Jurassic Kayenta Formation of Arizona. It is anatomically very similar to modern frogs. The oldest known caecilian is another Early Jurassic species, *Eocaecilia micropodia*, also from Arizona. The earliest salamander is *Beiyanerpeton jianpingensis* from the Late Jurassic of northeastern China.

Authorities disagree as to whether Salientia is a superorder that includes the order Anura, or whether Anura is a sub-order of the order Salientia. The Lissamphibia are traditionally divided into three orders, but an extinct salamander-like family, the Albanerpetontidae, is now considered part of Lissamphibia alongside the superorder Salientia. Furthermore, Salientia includes all three recent orders plus the Triassic proto-frog, *Triadobatrachus*.

Evolutionary History

Top: Restoration of *Eusthenopteron*, a fully aquatic lobe-finned fish
Bottom: Restoration of *Tiktaalik*, an advanced tetrapodomorph fish

The first major groups of amphibians developed in the Devonian period, around 370 million years ago, from lobe-finned fish which were similar to the modern coelacanth and lungfish. These ancient lobe-finned fish had evolved multi-jointed leg-like fins with digits that enabled them to crawl along the sea bottom. Some fish had developed primitive lungs to help them breathe air when the stagnant pools of the Devonian swamps were low in oxygen. They could also use their strong fins to hoist themselves out of the water and onto dry land if circumstances so required. Eventually, their bony fins would evolve into limbs and they would become the ancestors to all tetrapods, including modern amphibians, reptiles, birds, and mammals. Despite being able to crawl on land, many of these prehistoric tetrapodomorph fish still spent most of their time in the water. They had started to develop lungs, but still breathed predominantly with gills.

Many examples of species showing transitional features have been discovered. *Ichthyostega* was one of the first primitive amphibians, with nostrils and more efficient lungs. It had four sturdy limbs, a neck, a tail with fins and a skull very similar to that of the lobe-finned fish, *Eusthenopteron*. Amphibians evolved adaptations that allowed them to stay out of the water for longer periods. Their lungs improved and their skeletons became heavier and stronger, better able to support the weight of their bodies on land. They developed "hands" and "feet" with five or more digits; the skin became more capable of retaining body fluids and resisting desiccation. The fish's hyomandibula bone in the hyoid region behind the gills diminished in size and became the stapes of the amphibian ear, an adaptation necessary for hearing on dry land. An affinity between the amphibians and the teleost fish is the multi-folded structure of the teeth and the paired supra-occipital bones at the back of the head, neither of these features being found elsewhere in the animal kingdom.

The Permian lepospondyl *Diplocaulus* was largely aquatic

At the end of the Devonian period (360 million years ago), the seas, rivers and lakes were teeming with life while the land was the realm of early plants and devoid of vertebrates, though some, such as *Ichthyostega*, may have sometimes hauled themselves out of the water. It is thought they may have propelled themselves with their forelimbs, dragging their hindquarters in a similar manner to that used by the elephant seal. In the early Carboniferous (360 to 345 million years ago), the climate became wet and warm. Extensive swamps developed with mosses, ferns, horsetails and calamites. Air-breathing arthropods evolved and invaded the land where they provided food for the carnivorous amphibians that began to adapt to the terrestrial environment. There were no other tetrapods on the land and the amphibians were at the top of the food chain, occupying the ecological position currently held by the crocodile. Though equipped with limbs and the ability to breathe air, most still had a long tapering body and strong tail. They were the top land predators, sometimes reaching several metres in length, preying on the large insects of the period and the many types of fish in the water. They still needed to return to water to lay their shell-less eggs, and even most modern amphibians have a fully aquatic larval stage with gills like their fish ancestors. It was the development of the amniotic egg, which prevents the developing embryo from drying out, that enabled the reptiles to reproduce on land and which led to their dominance in the period that followed.

After the Carboniferous rainforest collapse amphibian dominance gave way to reptiles, and amphibians were further devastated by the Permian–Triassic extinction event. During the Triassic Period (250 to 200 million years ago), the reptiles continued to out-compete the amphibians, leading to a reduction in both the amphibians' size and their importance in the biosphere. According to the fossil record, Lissamphibia, which includes all modern amphibians and is the only surviving lineage, may have branched off from the extinct groups Temnospondyli and Lepospondyli at some period between the Late Carboniferous and the Early Triassic. The relative scarcity of fossil evidence precludes precise dating, but the most recent molecular study, based on multilocus sequence typing, suggests a Late Carboniferous/Early Permian origin for extant amphibians.

The temnospondyl *Eryops* had sturdy limbs to support its body on land

The origins and evolutionary relationships between the three main groups of amphibians is a matter of debate. A 2005 molecular phylogeny, based on rDNA analysis, suggests that salamanders and caecilians are more closely related to each other than they are to frogs. It also appears that the divergence of the three groups took place in the Paleozoic or early Mesozoic (around 250 million years ago), before the breakup of the supercontinent Pangaea and soon after their divergence from the lobe-finned fish. The briefness of this period, and the swiftness with which radiation took place, would help account for the relative scarcity of primitive amphibian fossils. There are large gaps in the fossil record, but the discovery of a proto-frog from the Early Permian in Texas in 2008 provided a missing link with many of the characteristics of modern frogs. Molecular analysis suggests that the frog–salamander divergence took place considerably earlier than the palaeontological evidence indicates. Newer research indicates that the common ancestor of all Lissamphibians lived about 315 million years ago, and that stereospondyls are the closest relatives to the caecilians.

As they evolved from lunged fish, amphibians had to make certain adaptations for living on land, including the need to develop new means of locomotion. In the water, the sideways thrusts of their tails had propelled them forward, but on land, quite different mechanisms were required. Their vertebral columns, limbs, limb girdles and musculature needed to be strong enough to raise them off the ground for locomotion and feeding. Terrestrial adults discarded their lateral line systems and adapted their sensory systems to receive stimuli via the medium of the air. They needed to develop new methods to regulate their body heat to cope with fluctuations in ambient temperature. They developed behaviours suitable for reproduction in a terrestrial environment. Their skins were exposed to harmful ultraviolet rays that had previously been absorbed by the water. The skin changed to become more protective and prevent excessive water loss.

Characteristics

The superclass Tetrapoda is divided into four classes of vertebrate animals with four limbs. Reptiles, birds and mammals are amniotes, the eggs of which are either laid or carried by the female and are surrounded by several membranes, some of which are impervious. Lacking these membranes, amphibians require water bodies for reproduction, although some species have developed various strategies for protecting or bypassing the vulnerable aquatic larval stage. They are not found in the sea with the exception of one or two frogs that live in brackish water in mangrove swamps. On land, amphibians are restricted to moist habitats because of the need to keep their skin damp.

The smallest amphibian (and vertebrate) in the world is a microhylid frog from New Guinea (*Paedophryne amauensis*) first discovered in 2012. It has an average length of 7.7 mm (0.30 in) and is part of a genus that contains four of the world's ten smallest frog species. The largest living amphibian is the 1.8 m (5 ft 11 in) Chinese giant salamander (*Andrias davidianus*) but this is a great deal smaller than the largest amphibian that ever existed—the extinct 9 m (30 ft) *Prionosuchus*, a crocodile-like temnospondyl dating to 270 million years ago from the middle Permian of Brazil! The largest frog is the African Goliath frog (*Conraua goliath*), which can reach 32 cm (13 in) and weigh 3 kg (6.6 lb).

Amphibians are ectothermic (cold-blooded) vertebrates that do not maintain their body temperature through internal physiological processes. Their metabolic rate is low and as a result, their food and energy requirements are limited. In the adult state, they have tear ducts and movable eyelids, and most species have ears that can detect airborne or ground vibrations. They have muscular tongues, which in many species can be protruded. Modern amphibians have fully ossified vertebrae with articular processes. Their ribs are usually short and may be fused to the vertebrae. Their skulls are mostly broad and short, and are often incompletely ossified. Their skin contains little keratin and lacks scales, apart from a few fish-like scales in certain caecilians. The skin contains many mucous glands and in some species, poison glands (a type of granular gland). The hearts of amphibians have three chambers, two atria and one ventricle. They have a urinary bladder and nitrogenous waste products are excreted primarily as urea. Most amphibians lay their eggs in water and have aquatic larvae that undergo metamorphosis to become terrestrial adults. Amphibians breathe by means of a pump action in which air is first drawn into the buccopharyngeal region through the nostrils. These are then closed and the air is forced into the lungs by contraction of the throat. They supplement this with gas exchange through the skin.

Anura

The order Anura (from the Ancient Greek *a(n)*- meaning "without" and *oura* meaning "tail") comprises the frogs and toads. They usually have long hind limbs that fold underneath them, shorter forelimbs, webbed toes with no claws, no tails, large eyes and glandular moist skin. Members of this order with smooth skins are commonly referred to as frogs, while those with warty skins are known as toads. The difference is not a formal one taxonomically and there are numerous exceptions to this rule. Members of the family Bufonidae are known as the "true toads". Frogs range in size from the 30-centimetre (12 in) Goliath frog (*Conraua goliath*) of West Africa to the 7.7-millimetre (0.30 in) *Paedophryne amauensis*, first described in Papua New Guinea in 2012, which is also the smallest known vertebrate. Although most species are associated with water and damp habitats, some are specialised to live in trees or in deserts. They are found worldwide except for polar areas.

Red-eyed tree frog (*Agalychnis callidryas*) with limbs and feet specialised for climbing

Anura is divided into three suborders that are broadly accepted by the scientific community, but the relationships between some families remain unclear. Future molecular studies should provide further insights into their evolutionary relationships. The suborder Archaeobatrachia contains four families of primitive frogs. These are Ascaphidae, Bombinatoridae, Discoglossidae and Leiopelmatidae which have few derived features and are probably paraphyletic with regard to other frog lineages. The six families in the more evolutionarily advanced suborder Mesobatrachia are the fossorial Megophryidae, Pelobatidae, Pelodytidae, Scaphiopodidae and Rhinophrynidae and the obligatorily aquatic Pipidae. These have certain characteristics that are intermediate between the two other suborders. Neobatrachia is by far the largest suborder and includes the remaining families of modern frogs, including most common species. Ninety-six percent of the over 5,000 extant species of frog are neobatrachians.

Caudata

Japanese giant salamander
(*Andrias japonicus*), a primitive salamander

The order Caudata (from the Latin *cauda* meaning "tail") consists of the salamanders—elongated, low-slung animals that mostly resemble lizards in form. This is a symplesiomorphic trait and they are no more closely related to lizards than they are to mammals. Salamanders lack claws, have scale-free skins, either smooth or covered with tubercles, and tails that are usually flattened from side to side and often finned. They range in size from the Chinese giant salamander (*Andrias davidianus*), which has been reported to grow to a length of 1.8 metres (5 ft 11 in), to the diminutive *Thorius pennatulus* from Mexico which seldom exceeds 20 mm (0.8 in) in length. Salamanders have a mostly Laurasian distribution, being present in much of the Holarctic region of the northern hemisphere. The family Plethodontidae is also found in Central America and South America north of the Amazon basin; South America was apparently invaded from Central America by about the start of the Miocene, 23 million years ago. Urodela is a name sometimes used for all the extant species of salamanders. Members of several salamander families have become paedomorphic and either fail to complete their metamorphosis or retain some larval characteristics as adults. Most salamanders are under 15 cm (6 in) long. They may be terrestrial or aquatic and many spend part of the year in each habitat. When on land, they mostly spend the day hidden under stones or logs or in dense vegetation, emerging in the evening and night to forage for worms, insects and other invertebrates.

Danube crested newt
(*Triturus dobrogicus*), an advanced salamander

The suborder Cryptobranchoidea contains the primitive salamanders. A number of fossil cryptobranchids have been found, but there are only three living species, the Chinese giant salamander (*Andrias davidianus*), the Japanese giant salamander (*Andrias japonicus*) and the hellbender (*Cryptobranchus alleganiensis*) from North America. These large amphibians retain several larval characteristics in their adult state; gills slits are present and the eyes are unlidded. A unique feature is their ability to feed by suction, depressing either the left side of their lower jaw or the right. The males excavate nests, persuade females to lay their egg strings inside them, and guard them. As well as breathing with lungs, they respire through the many folds in their thin skin, which has capillaries close to the surface.

The suborder Salamandroidea contains the advanced salamanders. They differ from the cryptobranchids by having fused prearticular bones in the lower jaw, and by using internal fertilisation. In salamandrids, the male deposits a bundle of sperm, the spermatophore, and the female picks it up and inserts it into her cloaca where the sperm is stored until the eggs are laid. The largest family in this group is Plethodontidae, the lungless salamanders, which includes 60% of all salamander species. The family Salamandridae includes the true salamanders and the name "newt" is given to members of its subfamily Pleurodelinae.

The third suborder, Sirenoidea, contains the four species of sirens, which are in a single family, Sirenidae. Members of this order are eel-like aquatic salamanders with much reduced forelimbs and no hind limbs. Some of their features are primitive while others are derived. Fertilisation is likely to be external as sirenids lack the cloacal glands used by male salamandrids to produce spermatophores and the females lack spermathecae for sperm storage. Despite this, the eggs are laid singly, a behaviour not conducive for external fertilisation.

Gymnophiona

The order Gymnophiona (from the Greek *gymnos* meaning "naked" and *ophis* meaning "serpent") or Apoda (from the Latin *an-* meaning "without" and the Greek *poda* meaning "legs") comprises the caecilians. These are long, cylindrical, limbless animals with a snake- or worm-like form. The adults vary in length from 8 to 75 centimetres (3 to 30 inches) with the exception of Thomson's caecilian (*Caecilia thompsoni*), which can reach 150 centimetres (4.9 feet). A caecilian's skin has a large number of transverse folds and in some species contains tiny embedded dermal scales. It has rudimentary eyes covered in skin, which are probably limited to discerning differences in light intensity. It also has a pair of short tentacles near the eye that can be extended and which have tactile and olfactory functions. Most caecilians live underground in burrows in damp soil, in rotten wood and under plant debris, but some are aquatic. Most species lay their eggs underground and when the larvae hatch, they make their way to adjacent bodies of water. Others brood their eggs and the larvae undergo metamorphosis before the eggs hatch. A few species give birth to live young, nourishing them with glandular secretions while they are in the oviduct. Caecilians have a mostly Gondwanan distribution, being found in tropical regions of Africa, Asia and Central and South America.

The limbless South American caecilian *Siphonops paulensis*

Anatomy and Physiology

Skin

The bright colours of the common reed frog (*Hyperolius viridiflavus*) are typical of a toxic species

The integumentary structure contains some typical characteristics common to terrestrial verte-brates, such as the presence of highly cornified outer layers, renewed periodically through a moult-ing process controlled by the pituitary and thyroid glands. Local thickenings (often called warts) are common, such as those found on toads. The outside of the skin is shed periodically mostly in one piece, in contrast to mammals and birds where it is shed in flakes. Amphibians often eat the sloughed skin. Caecilians are unique among amphibians in having mineralized dermal scales em-bedded in the dermis between the furrows in the skin. The similarity of these to the scales of bony fish is largely superficial. Lizards and some frogs have somewhat similar osteoderms forming bony deposits in the dermis, but this is an example of convergent evolution with similar structures hav-ing arisen independently in diverse vertebrate lineages.

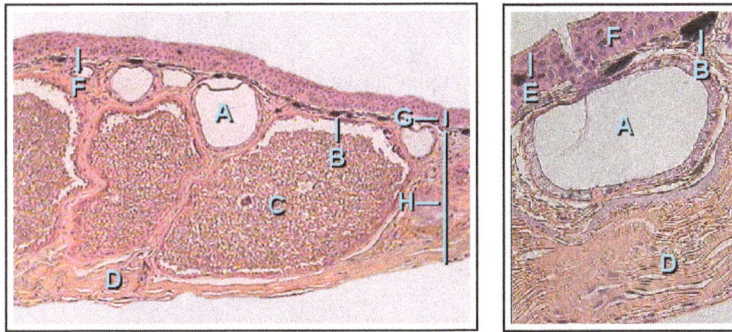

Cross section of frog skin. A: Mucus gland, B: Chromatophore, C: Granular poison gland, D: Connective tissue, E: Stratum corneum, F: Transition zone, G: Epidermis, H: Dermis

Amphibian skin is permeable to water. Gas exchange can take place through the skin (cutaneous respiration) and this allows adult amphibians to respire without rising to the surface of water and to hibernate at the bottom of ponds. To compensate for their thin and delicate skin, amphibians have evolved mucous glands, principally on their heads, backs and tails. The secretions produced by these help keep the skin moist. In addition, most species of amphibian have granular glands that secrete distasteful or poisonous substances. Some amphibian toxins can be lethal to humans while others have little effect. The main poison-producing glands, the paratoids, produce the neurotoxin bufotoxin and are located behind the ears of toads, along the backs of frogs, behind the eyes of salamanders and on the upper surface of caecilians.

The skin colour of amphibians is produced by three layers of pigment cells called chromatophores. These three cell layers consist of the melanophores (occupying the deepest layer), the guanophores (forming an intermediate layer and containing many granules, producing a blue-green colour) and the lipophores (yellow, the most superficial layer). The colour change displayed by many species is initiated by hormones secreted by the pituitary gland. Unlike bony fish, there is no direct control of the pigment cells by the nervous system, and this results in the colour change taking place more slowly than happens in fish. A vividly coloured skin usually indicates that the species is toxic and is a warning sign to predators.

Skeletal System and Locomotion

Amphibians have a skeletal system that is structurally homologous to other tetrapods, though with a number of variations. They all have four limbs except for the legless caecilians and a few species of salamander with reduced or no limbs. The bones are hollow and lightweight. The musculoskeletal system is strong to enable it to support the head and body. The bones are fully ossified and the vertebrae interlock with each other by means of overlapping processes. The pectoral girdle is supported by muscle, and the well-developed pelvic girdle is attached to the backbone by a pair of sacral ribs. The ilium slopes forward and the body is held closer to the ground than is the case in mammals.

In most amphibians, there are four digits on the fore foot and five on the hind foot, but no claws on either. Some salamanders have fewer digits and the amphiumas are eel-like in appearance with tiny, stubby legs. The sirens are aquatic salamanders with stumpy forelimbs and no hind limbs. The caecilians are limbless. They burrow in the manner of earthworms with zones of muscle contractions moving along the body. On the surface of the ground or in water they move by undulating their body from side to side.

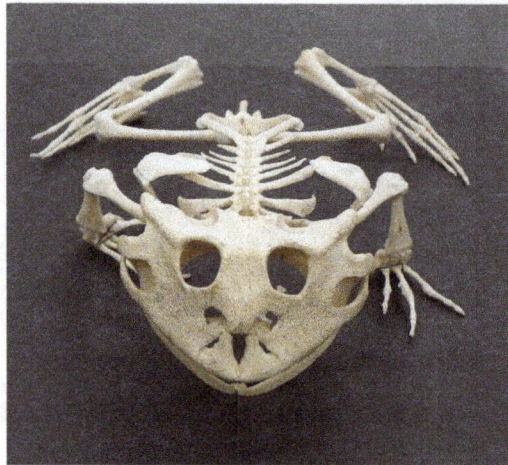

Skeleton of the Surinam horned frog
(*Ceratophrys cornuta*)

In frogs, the hind legs are larger than the fore legs, especially so in those species that principally move by jumping or swimming. In the walkers and runners the hind limbs are not so large, and the burrowers mostly have short limbs and broad bodies. The feet have adaptations for the way of life, with webbing between the toes for swimming, broad adhesive toe pads for climbing, and keratinised tubercles on the hind feet for digging (frogs usually dig backwards into the soil). In most salamanders, the limbs are short and more or less the same length and project at right angles from the body. Locomotion on land is by walking and the tail often swings from side to side or is used as a prop, particularly when climbing. In their normal gait, only one leg is advanced at a time in the manner adopted by their ancestors, the lobe-finned fish. Some salamanders in the genus *Aneides* and certain plethodontids climb trees and have long limbs, large toepads and prehensile tails. In aquatic salamanders and in frog tadpoles, the tail has dorsal and ventral fins and is moved from side to side as a means of propulsion. Adult frogs do not have tails and caecilians have only very short ones.

Salamanders use their tails in defence and some are prepared to jettison them to save their lives in a process known as autotomy. Certain species in the Plethodontidae have a weak zone at the base of the tail and use this strategy readily. The tail often continues to twitch after separation which may distract the attacker and allow the salamander to escape. Both tails and limbs can be regenerated. Adult frogs are unable to regrow limbs but tadpoles can do so.

Circulatory System

Amphibians have a juvenile stage and an adult stage, and the circulatory systems of the two are distinct. In the juvenile (or tadpole) stage, the circulation is similar to that of a fish; the two-chambered heart pumps the blood through the gills where it is oxygenated, and is spread around the body and back to the heart in a single loop. In the adult stage, amphibians (especially frogs) lose their gills and develop lungs. They have a heart that consists of a single ventricle and two atria. When the ventricle starts contracting, deoxygenated blood is pumped through the pulmonary artery to the lungs. Continued contraction then pumps oxygenated blood around the rest of the body. Mixing of the two bloodstreams is minimized by the anatomy of the chambers.

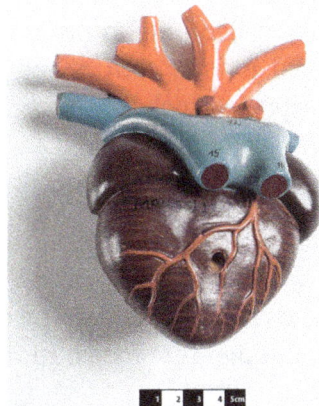

Didactic model of an amphibian heart.

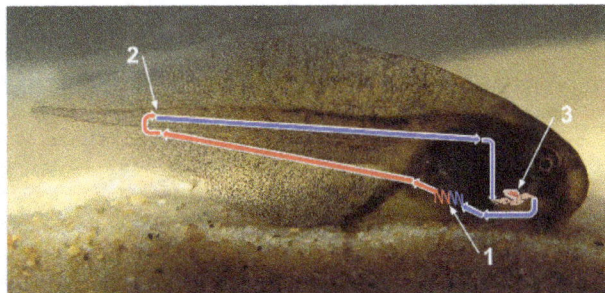

Juvenile amphibian circulatory systems are single loop systems which resemble fish.
1 – Internal gills where the blood is reoxygenated
2 – Point where the blood is depleted of oxygen and returns to the heart via veins
3 – Two chambered heart.
Red indicates oxygenated blood, and blue represents oxygen depleted blood.

Nervous and Sensory Systems

The nervous system is basically the same as in other vertebrates, with a central brain, a spinal cord, and nerves throughout the body. The amphibian brain is less well developed than that of reptiles, birds and mammals but is similar in morphology and function to that of a fish. It is believed amphibians are capable of perceiving pain. The brain consists of equal parts, cerebrum, midbrain and cerebellum. Various parts of the cerebrum process sensory input, such as smell in the olfactory lobe and sight in the optic lobe, and it is additionally the centre of behaviour and learning. The cerebellum is the center of muscular coordination and the medulla oblongata controls some organ functions including heartbeat and respiration. The brain sends signals through the spinal cord and nerves to regulate activity in the rest of the body. The pineal body, known to regulate sleep patterns in humans, is thought to produce the hormones involved in hibernation and aestivation in amphibians.

Tadpoles retain the lateral line system of their ancestral fishes, but this is lost in terrestrial adult amphibians. Some caecilians possess electroreceptors that allow them to locate objects around them when submerged in water. The ears are well developed in frogs. There is no external ear, but the large circular eardrum lies on the surface of the head just behind the eye. This vibrates and sound is transmitted through a single bone, the stapes, to the inner ear. Only high-frequency sounds like mating calls are heard in this way, but low-frequency noises can be detected through another mechanism. There is a patch of specialized haircells, called *papilla amphibiorum*, in the inner ear capable of detecting deeper sounds. Another feature, unique to frogs and salamanders, is the columella-operculum complex adjoining the auditory capsule which is involved in the transmission of both airborne and seismic signals. The ears of salamanders and caecilians are less highly developed than those of frogs as they do not normally communicate with each other through the medium of sound.

The eyes of tadpoles lack lids, but at metamorphosis, the cornea becomes more dome-shaped, the lens becomes flatter, and eyelids and associated glands and ducts develop. The adult eyes are an improvement on invertebrate eyes and were a first step in the development of more advanced vertebrate eyes. They allow colour vision and depth of focus. In the retinas are green rods, which are receptive to a wide range of wavelengths.

Digestive and Excretory Systems

Dissected frog: 1 Right atrium, 2 Liver, 3 Aorta, 4 Egg mass, 5 Colon, 6 Left atrium, 7 Ventricle, 8 Stomach, 9 Left lung, 10 Gallbladder, 11 Small intestine, 12 Cloaca

Many amphibians catch their prey by flicking out an elongated tongue with a sticky tip and drawing it back into the mouth before seizing the item with their jaws. Some use inertial feeding to help them swallow the prey, repeatedly thrusting their head forward sharply causing the food to move backwards in their mouth by inertia. Most amphibians swallow their prey whole without much chewing so they possess voluminous stomachs. The short oesophagus is lined with cilia that help to move the food to the stomach and mucus produced by glands in the mouth and pharynx eases its passage. The enzyme chitinase produced in the stomach helps digest the chitinous cuticle of arthropod prey.

Amphibians possess a pancreas, liver and gall bladder. The liver is usually large with two lobes. Its size is determined by its function as a glycogen and fat storage unit, and may change with the seasons as these reserves are built or used up. Adipose tissue is another important means of storing energy and this occurs in the abdomen (in internal structures called fat bodies), under the skin and, in some salamanders, in the tail.

There are two kidneys located dorsally, near the roof of the body cavity. Their job is to filter the blood of metabolic waste and transport the urine via ureters to the urinary bladder where it is stored before being passed out periodically through the cloacal vent. Larvae and most aquatic adult amphibians excrete the nitrogen as ammonia in large quantities of dilute urine, while terrestrial species, with a greater need to conserve water, excrete the less toxic product urea. Some tree frogs with limited access to water excrete most of their metabolic waste as uric acid.

Respiratory System

The axolotl (*Ambystoma mexicanum*) retains its larval form with gills into adulthood

The lungs in amphibians are primitive compared to those of amniotes, possessing few internal septa and large alveoli, and consequently having a comparatively slow diffusion rate for oxygen entering the blood. Ventilation is accomplished by buccal pumping. Most amphibians, however, are able to exchange gases with the water or air via their skin. To enable sufficient cutaneous respiration, the surface of their highly vascularised skin must remain moist to allow the oxygen to diffuse at a sufficiently high rate. Because oxygen concentration in the water increases at both low temperatures and high flow rates, aquatic amphibians in these situations can rely primarily on cutaneous respiration, as in the Titicaca water frog and the hellbender salamander. In air, where oxygen is more concentrated, some small species can rely solely on cutaneous gas exchange, most famously the plethodontid salamanders, which have neither lungs nor gills. Many aquatic salamanders and all tadpoles have gills in their larval stage, with some (such as the axolotl) retaining gills as aquatic adults.

Reproduction

For the purpose of reproduction most amphibians require fresh water although some lay their eggs on land and have developed various means of keeping them moist. A few (e.g. *Fejervarya raja*) can inhabit brackish water, but there are no true marine amphibians. There are reports, however, of particular amphibian populations unexpectedly invading marine waters. Such was the case with the Black Sea invasion of the natural hybrid *Pelophylax esculentus* reported in 2010.

Several hundred frog species in adaptive radiations (e.g., *Eleutherodactylus*, the Pacific *Platymantis*, the Australo-Papuan microhylids, and many other tropical frogs), however, do not need any water for breeding in the wild. They reproduce via direct development, an ecological and evolutionary adaptation that has allowed them to be completely independent from free-standing water. Almost all of these frogs live in wet tropical rainforests and their eggs hatch directly into miniature versions of the adult, passing through the tadpole stage within the egg. Reproductive success of many amphibians is dependent not only on the quantity of rainfall, but the seasonal timing.

Male orange-thighed frog (*Litoria xanthomera*) grasping the female during amplexus

In the tropics, many amphibians breed continuously or at any time of year. In temperate regions, breeding is mostly seasonal, usually in the spring, and is triggered by increasing day length, rising temperatures or rainfall. Experiments have shown the importance of temperature, but the trigger event, especially in arid regions, is often a storm. In anurans, males usually arrive at the breeding sites before females and the vocal chorus they produce may stimulate ovulation in females and the endocrine activity of males that are not yet reproductively active.

In caecilians, fertilisation is internal, the male extruding an intromittent organ, the phallodeum, and inserting it into the female cloaca. The paired Müllerian glands inside the male cloaca secrete a fluid which resembles that produced by mammalian prostate glands and which may transport and nourish the sperm. Fertilisation probably takes place in the oviduct.

The majority of salamanders also engage in internal fertilisation. In most of these, the male deposits a spermatophore, a small packet of sperm on top of a gelatinous cone, on the substrate either on land or in the water. The female takes up the sperm packet by grasping it with the lips of the cloaca and pushing it into the vent. The spermatozoa move to the spermatheca in the roof of the cloaca where they remain until ovulation which may be many months later. Courtship rituals and methods of transfer of the spermatophore vary between species. In some, the spermatophore may be placed directly into the female cloaca while in others, the female may be guided to the spermatophore or restrained with an embrace called amplexus. Certain primitive salamanders in the families Sirenidae, Hynobiidae and Cryptobranchidae practice external fertilisation in a similar manner to frogs, with the female laying the eggs in water and the male releasing sperm onto the egg mass.

With a few exceptions, frogs use external fertilisation. The male grasps the female tightly with his forelimbs either behind the arms or in front of the back legs, or in the case of *Epipedobates tricolor*, around the neck. They remain in amplexus with their cloacae positioned close together while the female lays the eggs and the male covers them with sperm. Roughened nuptial pads on the male's hands aid in retaining grip. Often the male collects and retains the egg mass, forming a sort of basket with the hind feet. An exception is the granular poison frog (*Oophaga granulifera*) where the male and female place their cloacae in close proximity while facing in opposite directions and then release eggs and sperm simultaneously. The tailed frog (*Ascaphus truei*) exhibits internal fertilisation. The "tail" is only possessed by the male and is an extension of the cloaca and used to inseminate the female. This frog lives in fast-flowing streams and internal fertilisation prevents the sperm from being washed away before fertilisation occurs. The sperm may be retained in storage tubes attached to the oviduct until the following spring.

Most frogs can be classified as either prolonged or explosive breeders. Typically, prolonged breeders congregate at a breeding site, the males usually arriving first, calling and setting up territories. Other satellite males remain quietly nearby, waiting for their opportunity to take over a territory. The females arrive sporadically, mate selection takes place and eggs are laid. The females depart and territories may change hands. More females appear and in due course, the breeding season comes to an end. Explosive breeders on the other hand are found where temporary pools appear in dry regions after rainfall. These frogs are typically fossorial species that emerge after heavy rains and congregate at a breeding site. They are attracted there by the calling of the first male to find a suitable place, perhaps a pool that forms in the same place each rainy season. The assembled frogs may call in unison and frenzied activity ensues, the males scrambling to mate with the usually smaller number of females.

Sexual selection has been studied in the red back salamander

There is a direct competition between males to win the attention of the females in salamanders and newts, with elaborate courtship displays to keep the female's attention long enough to get her interested in choosing him to mate with. Some species store sperm through long breeding seasons, as the extra time may allow for interactions with rival sperm.

Life Cycle

Most amphibians go through metamorphosis, a process of significant morphological change after birth. In typical amphibian development, eggs are laid in water and larvae are adapted to an aquatic lifestyle. Frogs, toads and salamanders all hatch from the egg as larvae with external gills. Metamorphosis in amphibians is regulated by thyroxine concentration in the blood, which stimulates metamorphosis, and prolactin, which counteracts thyroxine's effect. Specific events are dependent on threshold values for different tissues. Because most embryonic development is outside the parental body, it is subject to many adaptations due to specific environmental circumstances. For this reason tadpoles can have horny ridges instead of teeth, whisker-like skin extensions or fins. They also make use of a sensory lateral line organ similar to that of fish. After metamorphosis, these organs become redundant and will be reabsorbed by controlled cell death, called apoptosis. The variety of adaptations to specific environmental circumstances among amphibians is wide, with many discoveries still being made.

Eggs

The egg of an amphibian is typically surrounded by a transparent gelatinous covering secreted by the oviducts and containing mucoproteins and mucopolysaccharides. This capsule is permeable to water and gases, and swells considerably as it absorbs water. The ovum is at first rigidly held, but in fertilised eggs the innermost layer liquefies and allows the embryo to move freely. This also happens in salamander eggs, even when they are unfertilised. Eggs of some salamanders and frogs contain unicellular green algae. These penetrate the jelly envelope after the eggs are laid and may increase the supply of oxygen to the embryo through photosynthesis. They seem to both speed up the development of the larvae and reduce mortality. Most eggs contain the pigment melanin which raises their temperature through the absorption of light and also protects them against ultraviolet radiation. Caecilians, some plethodontid salamanders and certain frogs lay eggs underground that are unpigmented. In the wood frog (*Rana sylvatica*), the interior of the globular egg cluster has been found to be up to 6 °C (11 °F) warmer than its surroundings, which is an advantage in its cool northern habitat.

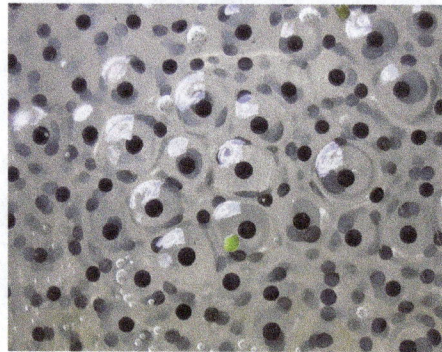

Frogspawn, a mass of eggs surrounded by jelly

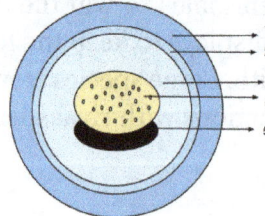

Amphibian egg:
1. Jelly capsule   2. Vitelline membrane   3. Perivitelline fluid   4. Yolk plug   5. Embryo

The eggs may be deposited singly or in small groups, or may take the form of spherical egg masses, rafts or long strings. In terrestrial caecilians, the eggs are laid in grape-like clusters in burrows near streams. The amphibious salamander *Ensatina* attaches its similar clusters by stalks to underwater stems and roots. The greenhouse frog (*Eleutherodactylus planirostris*) lays eggs in small groups in the soil where they develop in about two weeks directly into juvenile frogs without an intervening larval stage. The tungara frog (*Physalaemus pustulosus*) builds a floating nest from foam to protect its eggs. First a raft is built, then eggs are laid in the centre, and finally a foam cap is overlaid. The foam has anti-microbial properties. It contains no detergents but is created by whipping up proteins and lectins secreted by the female.

Larvae

The eggs of amphibians are typically laid in water and hatch into free-living larvae that complete their development in water and later transform into either aquatic or terrestrial adults. In many species of frog and in most lungless salamanders (Plethodontidae), direct development takes place, the larvae growing within the eggs and emerging as miniature adults. Many caecilians and some other amphibians lay their eggs on land, and the newly hatched larvae wriggle or are transported to water bodies. Some caecilians, the alpine salamander (*Salamandra atra*) and some of the African live-bearing toads (*Nectophrynoides spp.*) are viviparous. Their larvae feed on glandular secretions and develop within the female's oviduct, often for long periods. Other amphibians, but not caecilians, are ovoviviparous. The eggs are retained in or on the parent's body, but the larvae subsist on the yolks of their eggs and receive no nourishment from the adult. The larvae emerge at varying stages of their growth, either before or after metamorphosis, according to their species. The toad genus *Nectophrynoides* exhibits all of these developmental patterns among its dozen or so members.

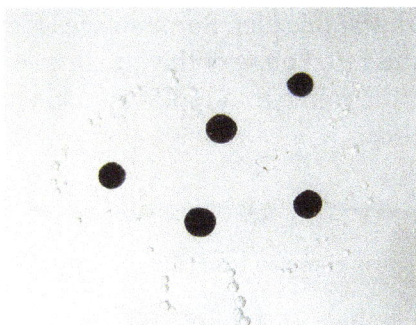

Early stages in the development of the embryos of the common frog (*Rana temporaria*)

Frogs

Frog larvae are known as tadpoles and typically have oval bodies and long, vertically flattened tails with fins. The free-living larvae are normally fully aquatic, but the tadpoles of some species (such as *Nannophrys ceylonensis*) are semi-terrestrial and live among wet rocks. Tadpoles have cartilaginous skeletons, gills for respiration (external gills at first, internal gills later), lateral line systems and large tails that they use for swimming. Newly hatched tadpoles soon develop gill pouches that cover the gills. The lungs develop early and are used as accessory breathing organs, the tadpoles rising to the water surface to gulp air. Some species complete their development inside the egg and hatch directly into small frogs. These larvae do not have gills but instead have specialised areas of skin through which respiration takes place. While tadpoles do not have true teeth, in most species, the jaws have long, parallel rows of small keratinized structures called keradonts surrounded by a horny beak. Front legs are formed under the gill sac and hind legs become visible a few days later.

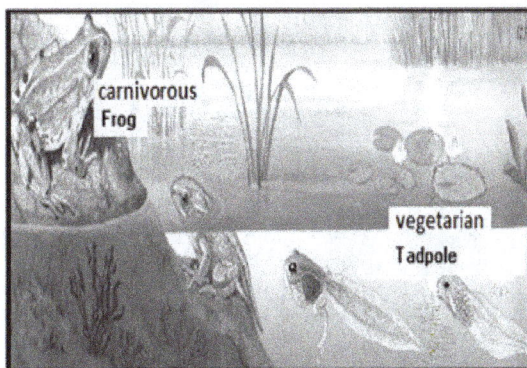

Amphibian metamorphosis

Iodine and T4 (over stimulate the spectacular apoptosis [programmed cell death] of the cells of the larval gills, tail and fins) also stimulate the evolution of nervous systems transforming the aquatic, vegetarian tadpole into the terrestrial, carnivorous frog with better neurological, visuospatial, olfactory and cognitive abilities for hunting.

In fact, tadpoles developing in ponds and streams are typically herbivorous. Pond tadpoles tend to have deep bodies, large caudal fins and small mouths; they swim in the quiet waters feeding on growing or loose fragments of vegetation. Stream dwellers mostly have larger mouths, shallow bodies and caudal fins; they attach themselves to plants and stones and feed on the surface films of algae and bacteria. They also feed on diatoms, filtered from the water through the gills, and stir up the sediment at bottom of the pond, ingesting edible fragments. They have a relatively long, spiral-shaped gut to enable them to digest this diet. Some species are carnivorous at the tadpole stage, eating insects, smaller tadpoles and fish. Young of the Cuban tree frog (*Osteopilus septentrionalis*) can occasionally be cannibalistic, the younger tadpoles attacking a larger, more developed tadpole when it is undergoing metamorphosis.

Successive stages in the development of common toad (*Bufo bufo*) tadpoles, finishing with metamorphosis

At metamorphosis, rapid changes in the body take place as the lifestyle of the frog changes completely. The spiral shaped mouth with horny tooth ridges is reabsorbed together with the spiral gut. The animal develops a large jaw, and its gills disappear along with its gill sac. Eyes and legs grow quickly, and a tongue is formed. There are associated changes in the neural networks such as development of stereoscopic vision and loss of the lateral line system. All this can happen in about a day. A few days later, the tail is reabsorbed, due to the higher thyroxine concentration required for this to take place.

Salamanders

Lungless salamanders in the family Plethodontidae are terrestrial and lay a small number of unpigmented eggs in a cluster among damp leaf litter. Each egg has a large yolk sac and the larva feeds on this while it develops inside the egg, emerging fully formed as a juvenile salamander. The female salamander often broods the eggs. In the genus *Ensatinas*, the female has been observed to coil around them and press her throat area against them, effectively massaging them with a mucous secretion.

Larva of the long-toed salamander (*Ambystoma macrodactylum*)

Larvae of the alpine newt (*Ichthyosaura alpestris*)

At hatching, a typical salamander larva has eyes without lids, teeth in both upper and lower jaws, three pairs of feathery external gills, a somewhat laterally flattened body and a long tail with dorsal and ventral fins. The forelimbs may be partially developed and the hind limbs are rudimentary in pond-living species but may be rather more developed in species that reproduce in moving water. Pond-type larvae often have a pair of balancers, rod-like structures on either side of the head that may prevent the gills from becoming clogged up with sediment. Some members of the genera *Ambystoma* and *Dicamptodon* have larvae that never fully develop into the adult form, but this varies with species and with populations. The northwestern salamander (*Ambystoma gracile*) is one of these and, depending on environmental factors, either remains permanently in the larval state, a condition known as neoteny, or transforms into an adult. Both of these are able to breed. Neoteny occurs when the animal's growth rate is very low and is usually linked to adverse conditions such as low water temperatures that may change the response of the tissues to the hormone thyroxine. Other factors that may inhibit metamorphosis include lack of food, lack of trace elements and competition from conspecifics. The tiger salamander (*Ambystoma tigrinum*) also sometimes behaves in this way and may grow particularly large in the process. The adult tiger salamander is terrestrial, but the larva is aquatic and able to breed while still in the larval state. When conditions are particularly inhospitable on land, larval breeding may allow continuation of a population that would otherwise die out. There are fifteen species of obligate neotenic salamanders, including species of *Necturus*, *Proteus* and *Amphiuma*, and many examples of facultative ones that adopt this strategy under appropriate environmental circumstances.

In newts and salamanders, metamorphosis is less dramatic than in frogs. This is because the larvae are already carnivorous and continue to feed as predators when they are adults so few changes are needed to their digestive systems. Their lungs are functional early, but the larvae do not make as much use of them as do tadpoles. Their gills are never covered by gill sacs and are reabsorbed just before the animals leave the water. Other changes include the reduction in size or loss of tail fins, the closure of gill slits, thickening of the skin, the development of eyelids, and certain changes in dentition and tongue structure. Salamanders are at their most vulnerable at metamorphosis as swimming speeds are reduced and transforming tails are encumbrances on land. Adult salamanders often have an aquatic phase in spring and summer, and a land phase in winter. For adaptation to a water phase, prolactin is the required hormone, and for adaptation to the land phase, thyroxine. External gills do not return in subsequent aquatic phases because these are completely absorbed upon leaving the water for the first time.

Caecilians

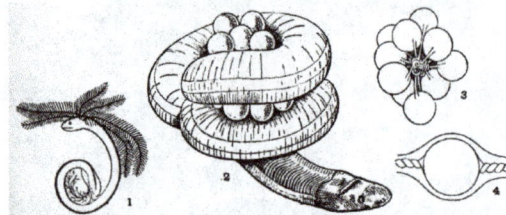

FIG. 15.—*Ichthyophis glutinosa* × 1.   (After P. and F. Sarasin.) 1, A nearly ripe embryo, with gills, tail-fin, and still with a considerable amount of yolk ; 2, female guarding her eggs, coiled up in a hole underground ; 3, a bunch of newly laid eggs ; 4, a single egg, enlarged, schematised to show the twisted albuminous strings or chalazae within the outer membrane, which surrounds the white of the egg.

The caecilian *Ichthyophis glutinosus* with eggs and developing embryo

Most terrestrial caecilians that lay eggs do so in burrows or moist places on land near bodies of water. The development of the young of *Ichthyophis glutinosus*, a species from Sri Lanka, has been much studied. The eel-like larvae hatch out of the eggs and make their way to water. They have three pairs of external red feathery gills, a blunt head with two rudimentary eyes, a lateral line system and a short tail with fins. They swim by undulating their body from side to side. They are mostly active at night, soon lose their gills and make sorties onto land. Metamorphosis is gradual. By the age of about ten months they have developed a pointed head with sensory tentacles near the mouth and lost their eyes, lateral line systems and tails. The skin thickens, embedded scales develop and the body divides into segments. By this time, the caecilian has constructed a burrow and is living on land.

The ringed caecilian (*Siphonops annulatus*) resembles an earthworm

In the majority of species of caecilians, the young are produced by vivipary. *Typhlonectes compressicauda*, a species from South America, is typical of these. Up to nine larvae can develop in the oviduct at any one time. They are elongated and have paired sac-like gills, small eyes and specialised scraping teeth. At first, they feed on the yolks of the eggs, but as this source of nourishment declines they begin to rasp at the ciliated epithelial cells that line the oviduct. This stimulates the secretion of fluids rich in lipids and mucoproteins on which they feed along with scrapings from the oviduct wall. They may increase their length sixfold and be two-fifths as long as their mother before being born. By this time they have undergone metamorphosis, lost their eyes and gills, developed a thicker skin and mouth tentacles, and reabsorbed their teeth. A permanent set of teeth grow through soon after birth.

The ringed caecilian (*Siphonops annulatus*) has developed a unique adaptation for the purposes of reproduction. The progeny feed on a skin layer that is specially developed by the adult in a phenomenon known as maternal dermatophagy. The brood feed as a batch for about seven minutes at intervals of approximately three days which gives the skin an opportunity to regenerate. Meanwhile, they have been observed to ingest fluid exuded from the maternal cloaca.

Parental care

Male common rocket frog (*Colostethus panamensis*) carrying tadpoles on his back

The care of offspring among amphibians has been little studied but, in general, the larger the number of eggs in a batch, the less likely it is that any degree of parental care takes place. Nevertheless, it is estimated that in up to 20% of amphibian species, one or both adults play some role in the care of the young. Those species that breed in smaller water bodies or other specialised habitats tend to have complex patterns of behaviour in the care of their young.

Many woodland salamanders lay clutches of eggs under dead logs or stones on land. The black mountain salamander (*Desmognathus welteri*) does this, the mother brooding the eggs and guarding them from predation as the embryos feed on the yolks of their eggs. When fully developed, they break their way out of the egg capsules and disperse as juvenile salamanders. The male hellbender, a primitive salamander, excavates an underwater nest and encourages females to lay there. The male then guards the site for the two or three months before the eggs hatch, using body undulations to fan the eggs and increase their supply of oxygen.

Male common midwife toad (*Alytes obstetricans*) carrying eggs

The male *Colostethus subpunctatus*, a tiny frog, protects the egg cluster which is hidden under a stone or log. When the eggs hatch, the male transports the tadpoles on his back, stuck there by a mucous secretion, to a temporary pool where he dips himself into the water and the tadpoles drop off. The male midwife toad (*Alytes obstetricans*) winds egg strings round his thighs and carries the eggs around for up to eight weeks. He keeps them moist and when they are ready to hatch, he visits a pond or ditch and releases the tadpoles. The female gastric-brooding frog (*Rheobatrachus spp.*) reared larvae in her stomach after swallowing either the eggs or hatchlings; however, this stage was never observed before the species became extinct. The tadpoles secrete a hormone that inhibits digestion in the mother whilst they develop by consuming their very large yolk supply. The pouched frog (*Assa darlingtoni*) lays eggs on the ground. When they hatch, the male carries the tadpoles around in brood pouches on his hind legs. The aquatic Surinam toad (*Pipa pipa*) raises its young in pores on its back where they remain until metamorphosis. The granular poison frog (*Oophaga granulifera*) is typical of a number of tree frogs in the poison dart frog family Dendrobatidae. Its eggs are laid on the forest floor and when they hatch, the tadpoles are carried one by one on the back of an adult to a suitable water-filled crevice such as the axil of a leaf or the rosette of a bromeliad. The female visits the nursery sites regularly and deposits unfertilised eggs in the water and these are consumed by the tadpoles.

Feeding and Diet

With a few exceptions, adult amphibians are predators, feeding on virtually anything that moves that they can swallow. The diet mostly consists of small prey that do not move too fast such as beetles, caterpillars, earthworms and spiders. The sirens (*Siren spp.*) often ingest aquatic plant material with the invertebrates on which they feed and a Brazilian tree frog (*Xenohyla truncata*) includes a large quantity of fruit in its diet. The Mexican burrowing toad (*Rhinophrynus dorsalis*) has a specially adapted tongue for picking up ants and termites. It projects it with the tip foremost whereas other frogs flick out the rear part first, their tongues being hinged at the front.

Food is mostly selected by sight, even in conditions of dim light. Movement of the prey triggers a feeding response. Frogs have been caught on fish hooks baited with red flannel and green frogs (*Rana clamitans*) have been found with stomachs full of elm seeds that they had seen floating past. Toads, salamanders and caecilians also use smell to detect prey. This response is mostly secondary because salamanders have been observed to remain stationary near odoriferous prey but only feed if it moves. Cave-dwelling amphibians normally hunt by smell. Some salamanders seem to have learned to recognize immobile prey when it has no smell, even in complete darkness.

Northwestern salamander
(*Ambystoma gracile*) eating a worm

Amphibians usually swallow food whole but may chew it lightly first to subdue it. They typically have small hinged pedicellate teeth, a feature unique to amphibians. The base and crown of these are composed of dentine separated by an uncalcified layer and they are replaced at intervals. Salamanders, caecilians and some frogs have one or two rows of teeth in both jaws, but some frogs (*Rana spp.*) lack teeth in the lower jaw, and toads (*Bufo spp.*) have no teeth. In many amphibians there are also vomerine teeth attached to a facial bone in the roof of the mouth.

Edible frog (*Pelophylax esculentus*) exhibiting cannibalism

The tiger salamander (*Ambystoma tigrinum*) is typical of the frogs and salamanders that hide under cover ready to ambush unwary invertebrates. Others amphibians, such as the *Bufo spp.* toads, actively search for prey, while the Argentine horned frog (*Ceratophrys ornata*) lures inquisitive prey closer by raising its hind feet over its back and vibrating its yellow toes. Among leaf litter frogs in Panama, frogs that actively hunt prey have narrow mouths and are slim, often brightly coloured and toxic, while ambushers have wide mouths and are broad and well-camouflaged. Caecilians do not flick their tongues, but catch their prey by grabbing it with their slightly backward-pointing teeth. The struggles of the prey and further jaw movements work it inwards and the caecilian usually retreats into its burrow. The subdued prey is gulped down whole.

When they are newly hatched, frog larvae feed on the yolk of the egg. When this is exhausted some move on to feed on bacteria, algal crusts, detritus and raspings from submerged plants. Water is drawn in through their mouths, which are usually at the bottom of their heads, and passes through branchial food traps between their mouths and their gills where fine particles are trapped in mucus and filtered out. Others have specialised mouthparts consisting of a horny beak edged by several rows of labial teeth. They scrape and bite food of many kinds as well as stirring up the bottom sediment, filtering out larger particles with the papillae around their mouths. Some, such as the spadefoot toads, have strong biting jaws and are carnivorous or even cannibalistic.

Vocalization

Male treefrog (*Dendropsophus microcephalus*) inflating his air sac as he calls

The calls made by caecilians and salamanders are limited to occasional soft squeaks, grunts or hisses and have not been much studied. A clicking sound sometimes produced by caecilians may be a means of orientation, as in bats, or a form of communication. Most salamanders are considered voiceless, but the California giant salamander (*Dicamptodon ensatus*) has vocal cords and can produce a rattling or barking sound. Some species of salamander emit a quiet squeak or yelp if attacked.

American toad, (*Anaxyrus americanus*) singing

Frogs are much more vocal, especially during the breeding season when they use their voices to attract mates. The presence of a particular species in an area may be more easily discerned by its characteristic call than by a fleeting glimpse of the animal itself. In most species, the sound is produced by expelling air from the lungs over the vocal cords into an air sac or sacs in the throat or at the corner of the mouth. This may distend like a balloon and acts as a resonator, helping to transfer the sound to the atmosphere, or the water at times when the animal is submerged. The main vocalisation is the male's loud advertisement call which seeks to both encourage a female to approach and discourage other males from intruding on its territory. This call is modified to a quieter courtship call on the approach of a female or to a more aggressive version if a male intruder draws near. Calling carries the risk of attracting predators and involves the expenditure of much energy. Other calls include those given by a female in response to the advertisement call and a release call given by a male or female during unwanted attempts at amplexus. When a frog is attacked, a distress or fright call is emitted, often resembling a scream. The usually nocturnal Cuban tree frog (*Osteopilus septentrionalis*) produces a rain call when there is rainfall during daylight hours.

Territorial Behaviour

Little is known of the territorial behaviour of caecilians, but some frogs and salamanders defend home ranges. These are usually feeding, breeding or sheltering sites. Males normally exhibit such behaviour though in some species, females and even juveniles are also involved. Although in many frog species, females are larger than males, this is not the case in most species where males are actively involved in territorial defence. Some of these have specific adaptations such as enlarged teeth for biting or spines on the chest, arms or thumbs.

In salamanders, defence of a territory involves adopting an aggressive posture and if necessary attacking the intruder. This may involve snapping, chasing and sometimes biting, occasionally causing the loss of a tail. The behaviour of red back salamanders (*Plethodon cinereus*) has been much studied. 91% of marked individuals that were later recaptured were within a metre (yard) of their original daytime retreat under a log or rock. A similar proportion, when moved experimentally a distance of 30 metres (98 ft), found their way back to their home base. The salamanders left odour marks around their territories which averaged 0.16 to 0.33 square metres (1.7 to 3.6 sq ft) in size and were sometimes inhabited by a male and female pair. These deterred the intrusion of others and delineated the boundaries between neighbouring areas. Much of their behaviour seemed stereotyped and did not involve any actual contact between individuals. An aggressive posture involved raising the body off the ground and glaring at the opponent who often turned away submissively. If the intruder persisted, a biting lunge was usually launched at either the tail region or the naso-labial grooves. Damage to either of these areas can reduce the fitness of the rival, either because of the need to regenerate tissue or because it impairs its ability to detect food.

The red back salamander (*Plethodon cinereus*) defends a territory against intruders.

In frogs, male territorial behaviour is often observed at breeding locations; calling is both an announcement of ownership of part of this resource and an advertisement call to potential mates. In general, a deeper voice represents a heavier and more powerful individual, and this may be sufficient to prevent intrusion by smaller males. Much energy is used in the vocalization and it takes a toll on the territory holder who may be displaced by a fitter rival if he tires. There is a tendency for males to tolerate the holders of neighbouring territories while vigorously attacking unknown intruders. Holders of territories have a "home advantage" and usually come off better in an encounter between two similar-sized frogs. If threats are insufficient, chest to chest tussles may take place. Fighting methods include pushing and shoving, deflating the opponent's vocal sac, seizing him by the head, jumping on his back, biting, chasing, splashing, and ducking him under the water.

Defence Mechanisms

Cane toad (*Rhinella marina*) with poison glands behind the eyes

Amphibians have soft bodies with thin skins, and lack claws, defensive armour, or spines. Nevertheless, they have evolved various defence mechanisms to keep themselves alive. The first line of defence in salamanders and frogs is the mucous secretion that they produce. This keeps their skin moist and makes them slippery and difficult to grip. The secretion is often sticky and distasteful or toxic. Snakes have been observed yawning and gaping when trying to swallow African clawed frogs (*Xenopus laevis*), which gives the frogs an opportunity to escape. Caecilians have been little studied in this respect, but the Cayenne caecilian (*Typhlonectes compressicauda*) produces toxic mucus that has killed predatory fish in a feeding experiment in Brazil. In some salamanders, the skin is poisonous. The rough-skinned newt (*Taricha granulosa*) from North America and other members of its genus contain the neurotoxin tetrodotoxin (TTX), the most toxic non-protein substance known and almost identical to that produced by pufferfish. Handling the newts does not cause harm, but ingestion of even the most minute amounts of the skin is deadly. In feeding trials, fish, frogs, reptiles, birds and mammals were all found to be susceptible. The only predators with some tolerance to the poison are certain populations of common garter snake (*Thamnophis sirtalis*). In locations where both snake and salamander co-exist, the snakes have developed immunity through genetic changes and they feed on the amphibians with impunity. Coevolution occurs with the newt increasing its toxic capabilities at the same rate as the snake further develops its immunity. Some frogs and toads are toxic, the main poison glands being at the side of the neck and under the warts on the back. These regions are presented to the attacking animal and their secretions may be foul-tasting or cause various physical or neurological symptoms. Altogether, over 200 toxins have been isolated from the limited number of amphibian species that have been investigated.

The fire salamander (*Salamandra salamandra*), a toxic species, wears warning colours.

Poisonous species often use bright colouring to warn potential predators of their toxicity. These warning colours tend to be red or yellow combined with black, with the fire salamander (*Salamandra salamandra*) being an example. Once a predator has sampled one of these, it is likely to remember the colouration next time it encounters a similar animal. In some species, such as the fire-bellied toad (*Bombina spp.*), the warning colouration is on the belly and these animals adopt a defensive pose when attacked, exhibiting their bright colours to the predator. The frog *Allobates zaparo* is not poisonous, but mimics the appearance of other toxic species in its locality, a strategy that may deceive predators.

Many amphibians are nocturnal and hide during the day, thereby avoiding diurnal predators that hunt by sight. Other amphibians use camouflage to avoid being detected. They have various colourings such as mottled browns, greys and olives to blend into the background. Some salamanders adopt defensive poses when faced by a potential predator such as the North American northern short-tailed shrew (*Blarina brevicauda*). Their bodies writhe and they raise and lash their tails which makes it difficult for the predator to avoid contact with their poison-producing granular glands. A few salamanders will autotomise their tails when attacked, sacrificing this part of their anatomy to enable them to escape. The tail may have a constriction at its base to allow it to be easily detached. The tail is regenerated later, but the energy cost to the animal of replacing it is significant. Some frogs and toads inflate themselves to make themselves look large and fierce, and some spadefoot toads (*Pelobates spp*) scream and leap towards the attacker. Giant salamanders of the genus *Andrias*, as well as Ceratophrine and *Pyxicephalus* frogs possess sharp teeth and are capable of drawing blood with a defensive bite. The blackbelly salamander (*Desmognathus quadramaculatus*) can bite an attacking common garter snake (*Thamnophis sirtalis*) two or three times its size on the head and often manages to escape.

Perhaps the most poisonous animal in the world, the golden poison frog
(*Phyllobates terribilis*) is endemic to Colombia.

Cognition

In amphibians, there is evidence of habituation, associative learning through both classical and instrumental learning, and discrimination abilities.

In one experiment, when offered live fruit flies (*Drosophila virilis*), salamanders choose the larger of 1 vs 2 and 2 vs 3. Frogs can distinguish between low numbers (1 vs 2, 2 vs 3, but not 3 vs 4) and large numbers (3 vs 6, 4 vs 8, but not 4 vs 6) of prey. This is irrespective of other characteristics, i.e. surface area, volume, weight and movement, although discrimination among large numbers may be based on surface area.

Conservation

Dramatic declines in amphibian populations, including population crashes and mass localized extinction, have been noted since the late 1980s from locations all over the world, and amphibian declines are thus perceived to be one of the most critical threats to global biodiversity. In 2004, the International Union for Conservation of Nature (IUCN) reported stating that currently birds, mammals, and amphibians extinction rates were at minimum 48 times greater than natural extinction rates—possibly 1,024 times higher. In 2006 there were believed to be 4,035 species of amphibians that depended on water at some stage during their life cycle. Of these, 1,356 (33.6%) were considered to be threatened and this figure is likely to be an underestimate because it excludes 1,427 species for which there was insufficient data to assess their status. A number of causes are believed to be involved, including habitat destruction and modification, over-exploitation, pollution, introduced species, climate change, endocrine-disrupting pollutants, destruction of the ozone layer (ultraviolet radiation has shown to be especially damaging to the skin, eyes, and eggs of amphibians), and diseases like chytridiomycosis. However, many of the causes of amphibian declines are still poorly understood, and are a topic of ongoing discussion.

The extinct golden toad (*Bufo periglenes*), last seen in 1989

With their complex reproductive needs and permeable skins, amphibians are often considered to be ecological indicators. In many terrestrial ecosystems, they constitute one of the largest parts of the vertebrate biomass. Any decline in amphibian numbers will affect the patterns of predation. The loss of carnivorous species near the top of the food chain will upset the delicate ecosystem balance and may cause dramatic increases in opportunistic species. In the Middle East, a growing appetite for eating frog legs and the consequent gathering of them for food was linked to an increase in mosquitoes. Predators that feed on amphibians are affected by their decline. The western terrestrial garter snake (*Thamnophis elegans*) in California is largely aquatic and depends heavily on two species of frog that are diminishing in numbers, the Yosemite toad (*Bufo canorus*) and the mountain yellow-legged frog (*Rana muscosa*), putting the snake's future at risk. If the snake were to become scarce, this would affect birds of prey and other predators that feed on it. Meanwhile, in the ponds and lakes, fewer frogs means fewer tadpoles. These normally play an important role in controlling the growth of algae and also forage on detritus that accumulates as sediment on the bottom. A reduction in the number of tadpoles may lead to an overgrowth of algae, resulting in depletion of oxygen in the water when the algae later die and decompose. Aquatic invertebrates and fish might then die and there would be unpredictable ecological consequences.

The Hula painted frog (*Discoglossus nigriventer*) was believed to be extinct but was rediscovered in 2011.

A global strategy to stem the crisis was released in 2005 in the form of the Amphibian Conservation Action Plan. Developed by over eighty leading experts in the field, this call to action details what would be required to curtail amphibian declines and extinctions over the following five years and how much this would cost. The Amphibian Specialist Group of the IUCN is spearheading efforts to implement a comprehensive global strategy for amphibian conservation. Amphibian Ark is an organization that was formed to implement the ex-situ conservation recommendations of this plan, and they have been working with zoos and aquaria around the world, encouraging them to create assurance colonies of threatened amphibians. One such project is the Panama Amphibian Rescue and Conservation Project that built on existing conservation efforts in Panama to create a country-wide response to the threat of chytridiomycosis.

Invertebrate

Invertebrates are animals that neither possess nor develop a vertebral column (commonly known as a *backbone* or *spine*), derived from the notochord. This includes all animals apart from the subphylum Vertebrata. Familiar examples of invertebrates include insects; crabs, lobsters and their kin; snails, clams, octopuses and their kin; starfish, sea-urchins and their kin; jellyfish, and worms.

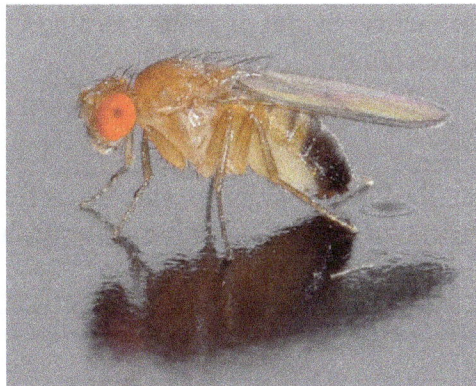

The common fruit fly, *Drosophila melanogaster*, has been used extensively for research.

The majority of animal species are invertebrates; one estimate puts the figure at 97%. Many invertebrate taxa have a greater number and variety of species than the entire subphylum of Vertebrata.

Some of the so-called invertebrates, such as the Tunicata and Cephalochordata are more closely related to the vertebrates than to other invertebrates. This makes the term "invertebrate" paraphyletic and hence almost meaningless for taxonomic purposes.

Etymology

The word "invertebrate" comes from the form of the Latin word *vertebra*, which means a joint in general, and sometimes specifically a joint from the spinal column of a vertebrate. In turn the jointed aspect of *vertebra* derived from the concept of turning, expressed in the root *verto* or *vorto*, to turn. Coupled with the prefix *in-*, meaning "not" or "without".

Taxonomic Significance

The term *invertebrates* is not always precise among non-biologists since it does not accurately describe a taxon in the same way that Arthropoda, Vertebrata or Manidae do. Each of these terms describes a valid taxon, phylum, subphylum or family. "Invertebrata" is a term of convenience, not a taxon; it has very little circumscriptional significance except within the Chordata. The Vertebrata as a subphylum comprises such a small proportion of the Metazoa that to speak of the kingdom Animalia in terms of "Vertebrata" and "Invertebrata" has limited practicality. In the more formal taxonomy of Animalia other attributes that logically should precede the presence or absence of the vertebral column in constructing a cladogram, for example, the presence of a notochord. That would at least circumscribe the Chordata. However, even the notochord would be a less fundamental criterion than aspects of embryological development and symmetry or perhaps bauplan.

Despite this, the concept of *invertebrates* as a taxon of animals has persisted for over a century among the laity, and within the zoological community and in its literature it remains in use as a term of convenience for animals that are not members of the Vertebrata. The following text reflects earlier scientific understanding of the term and of those animals which have constituted it. According to this understanding, invertebrates do not possess a skeleton of bone, either internal or external. They include hugely varied body plans. Many have fluid-filled, hydrostatic skeletons, like jellyfish or worms. Others have hard exoskeletons, outer shells like those of insects and crustaceans. The most familiar invertebrates include the Protozoa, Porifera, Coelenterata, Platyhelminthes, Nematoda, Annelida, Echinodermata, Mollusca and Arthropoda. Arthropoda include insects, crustaceans and arachnids.

Number of Extant Species

By far the largest number of described invertebrate species are insects. The following table lists the number of described extant species for major invertebrate groups as estimated in the IUCN Red List of Threatened Species, *2014.3*.

Invertebrate group	Image	Estimated number of described species
Insects		1,000,000
Molluscs		85,000
Crustaceans		47,000

Corals		2,000
Arachnids		102,248
Velvet worms		165
Horseshoe crabs		4
Others jellyfish, echinoderms, sponges, other worms etc.		68,658
	Totals:	**1,305,075**

The IUCN estimates that 66,178 extant vertebrate species have been described, which means that over 95% of the described animal species in the world are invertebrates.

Characteristics

The trait that is common to all invertebrates is the absence of a vertebral column (backbone): this creates a distinction between invertebrates and vertebrates. The distinction is one of convenience only; it is not based on any clear biologically homologous trait, any more than the common trait of having wings functionally unites insects, bats, and birds, or than not having wings unites tortoises, snails and sponges. Being animals, invertebrates are heterotrophs, and require sustenance in the form of the consumption of other organisms. With a few exceptions, such as the Porifera, invertebrates generally have bodies composed of differentiated tissues. There is also typically a digestive chamber with one or two openings to the exterior.

Morphology and Symmetry

The body plans of most multicellular organisms exhibit some form of symmetry, whether radial, bilateral, or spherical. A minority, however, exhibit no symmetry. One example of asymmetric invertebrates include all gastropod species. This is easily seen in snails and sea snails, which have helical shells. Slugs appear externally symmetrical, but their pneumostome (breathing hole) is located on the right side. Other gastropods develop external asymmetry, such as Glaucus atlanticus that develops asymmetrical cerata as they mature. The origin of gastropod asymmetry is a subject of scientific debate.

Other examples of asymmetry are found in fiddler crabs and hermit crabs. They often have one claw much larger than the other. If a male fiddler loses its large claw, it will grow another on the opposite side after moulting. Sessile animals such as sponges are asymmetrical alongside coral colonies (with the exception of the individual polyps that exhibit radial symmetry); alpheidae claws that lack pincers; and some copepods, polyopisthocotyleans, and monogeneans which parasitize by attachment or residency within the gill chamber of their fish hosts).

Nervous System

Neurons differ in invertebrates from mammalian cells. Invertebrates cells fire in response to similar stimuli as mammals, such as tissue trauma, high temperature, or changes in pH. The first invertebrate in which a neuron cell was identified was the medicinal leech, *Hirudo medicinalis*.

Learning and memory using nociceptors in the sea hare, *Aplysia* has been described. Mollusk neurons are able to detect increasing pressures and tissue trauma.

Neurons have been identified in a wide range of invertebrate species, including annelids, molluscs, nematodes and arthropods.

Respiratory System

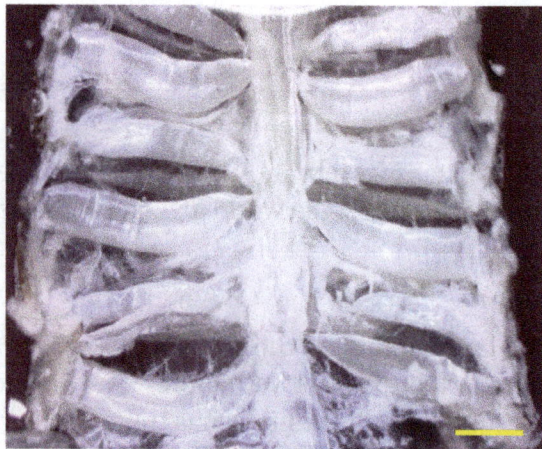

Tracheal system of dissected cockroach. The largest tracheae run across the width of the body of the cockroach and are horizontal in this image. Scale bar, 2 mm.

One type of invertebrate respiriatory system is the open respiratory system composed of spiracles, tracheae, and tracheoles that terrestrial arthropods have to transport metabolic gases to and from tissues. The distribution of spiracles can vary greatly among the many orders of insects, but in general each segment of the body can have only one pair of spiracles, each of which connects to an atrium and has a relatively large tracheal tube behind it. The tracheae are invaginations of the cuticular exoskeleton that branch (anastomose) throughout the body with diameters from only a few micrometres up to 0.8 mm. The smallest tubes, tracheoles, penetrate cells and serve as sites of diffusion for water, oxygen, and carbon dioxide. Gas may be conducted through the respiratory system by means of active ventilation or passive diffusion. Unlike vertebrates, insects do not generally carry oxygen in their haemolymph.

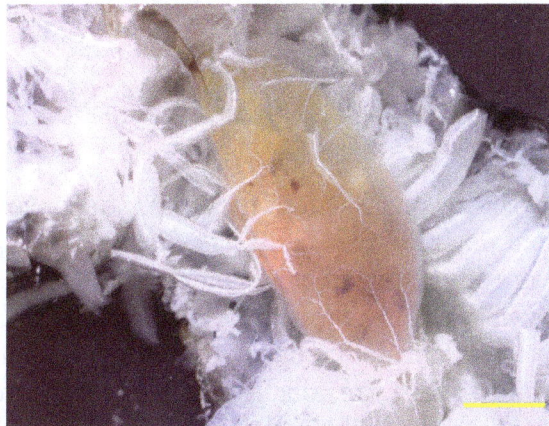

The tracheal system branches into progressively smaller tubes, here supplying the crop of the cockroach.
Scale bar, 2.0 mm.

A tracheal tube may contain ridge-like circumferential rings of taenidia in various geometries such as loops or helices. In the head, thorax, or abdomen, tracheae may also be connected to air sacs. Many insects, such as grasshoppers and bees, which actively pump the air sacs in their abdomen, are able to control the flow of air through their body. In some aquatic insects, the tracheae exchange gas through the body wall directly, in the form of a gill, or function essentially as normal, via a plastron. Note that despite being internal, the tracheae of arthropods are shed during moulting (ecdysis).

Reproduction

Like vertebrates, most invertebrates reproduce at least partly through sexual reproduction. They produce specialized reproductive cells that undergo meiosis to produce smaller, motile spermatozoa or larger, non-motile ova. These fuse to form zygotes, which develop into new individuals. Others are capable of asexual reproduction, or sometimes, both methods of reproduction.

Social Interaction

Social behavior is widespread in invertebrates, including cockroaches, termites, aphids, thrips, ants, bees, Passalidae, Acari, spiders, and more. Social interaction is particularly salient in eusocial species but applies to other invertebrates as well.

Insects recognize information transmitted by other insects.

Phyla

The fossil coral *Cladocora* from the Pliocene of Cyprus

The term invertebrates covers several phyla. One of these are the sponges (Porifera). They were long thought to have diverged from other animals early. They lack the complex organization found in most other phyla. Their cells are differentiated, but in most cases not organized into distinct tissues. Sponges typically feed by drawing in water through pores. Some speculate that sponges are not so primitive, but may instead be secondarily simplified. The Ctenophora and the Cnidaria, which includes sea anemones, corals, and jellyfish, are radially symmetric and have digestive chambers with a single opening, which serves as both the mouth and the anus. Both have distinct tissues, but they are not organized into organs. There are only two main germ layers, the ectoderm and endoderm, with only scattered cells between them. As such, they are sometimes called diploblastic.

The Echinodermata are radially symmetric and exclusively marine, including starfish (Asteroidea), sea urchins, (Echinoidea), brittle stars (Ophiuroidea), sea cucumbers (Holothuroidea) and feather stars (Crinoidea).

The largest animal phylum is also included within invertebrates is the Arthropoda, including insects, spiders, crabs, and their kin. All these organisms have a body divided into repeating segments, typically with paired appendages. In addition, they possess a hardened exoskeleton that is periodically shed during growth. Two smaller phyla, the Onychophora and Tardigrada, are close relatives of the arthropods and share these traits. The Nematoda or roundworms, are perhaps the second largest animal phylum, and are also invertebrates. Roundworms are typically microscopic, and occur in nearly every environment where there is water. A number are important parasites. Smaller phyla related to them are the Kinorhyncha, Priapulida, and Loricifera. These groups have a reduced coelom, called a pseudocoelom. Other invertebrates include the Nemertea or ribbon worms, and the Sipuncula.

Another phylum is Platyhelminthes, the flatworms. These were originally considered primitive, but it now appears they developed from more complex ancestors. Flatworms are acoelomates, lacking a body cavity, as are their closest relatives, the microscopic Gastrotricha. The Rotifera or rotifers, are common in aqueous environments. Invertebrates also include the Acanthocephala or spiny-headed worms, the Gnathostomulida, Micrognathozoa, and the Cycliophora.

Also included are two of the most successful animal phyla, the Mollusca and Annelida. The former, which is the second-largest animal phylum by number of described species, includes animals such as snails, clams, and squids, and the latter comprises the segmented worms, such as earthworms and leeches. These two groups have long been considered close relatives because of the common presence of trochophore larvae, but the annelids were considered closer to the arthropods because they are both segmented. Now, this is generally considered convergent evolution, owing to many morphological and genetic differences between the two phyla.

Among lesser phyla of invertebrates are the Hemichordata, or acorn worms, and the Chaetognatha, or arrow worms. Other phyla include Acoelomorpha, Brachiopoda, Bryozoa, Entoprocta, Phoronida, and Xenoturbellida.

Classification of Invertebrates

Invertebrates can be classified into several main categories, some of which are taxonomically obsolescent or debatable, but still used as terms of convenience. Each however appears in its own article at the following links.

- *Protozoa* (like the worms, an arbitrary grouping of convenience)
- Sponges (*Porifera*)
- Stinging jellyfish and corals (*Cnidaria*)
- Comb jellies (*Ctenophora*)
- Flatworms (*Platyhelminthes*)
- Round- or threadworms (*Nematoda*)
- segmented worms (*Annelida*)
- Insects, spiders, crabs and their kin (*Arthropoda*)
- Cuttlefish, snails, mussels and their kin (*Mollusca*)
- Starfish, sea-cucumbers and their kin (*Echinodermata*)

History

The earliest animal fossils appear to be those of invertebrates. 665-million-year-old fossils in the Trezona Formation at Trezona Bore, West Central Flinders, South Australia have been interpreted as being early sponges. Some paleontologists suggest that animals appeared much earlier, possibly as early as 1 billion years ago. Trace fossils such as tracks and burrows found in the Tonian era indicate the presence of triploblastic worms, like metazoans, roughly as large (about 5 mm wide) and complex as earthworms.

Around 453 MYA, animals began diversifying, and many of the important groups of invertebrates diverged from one another. Fossils of invertebrates are found in various types of sediment from the Phanerozoic. Fossils of invertebrates are commonly used in stratigraphy.

Classification

Carl Linnaeus divided these animals into only two groups, the Insecta and the now-obsolete Vermes (worms). Jean-Baptiste Lamarck, who was appointed to the position of "Curator of Insecta and Vermes" at the Muséum National d'Histoire Naturelle in 1793, both coined the term "invertebrate" to describe such animals, and divided the original two groups into ten, by splitting Arachnida and Crustacea from the Linnean Insecta, and Mollusca, Annelida, Cirripedia, Radiata, Coelenterata and Infusoria from the Linnean Vermes. They are now classified into over 30 phyla, from simple organisms such as sea sponges and flatworms to complex animals such as arthropods and molluscs.

Significance of the Group

Invertebrates are animals *without* a vertebral column. This has led to the conclusion that *in*vertebrates are a group that deviates from the normal, vertebrates. This has been said to be because researchers in the past, such as Lamarck, viewed vertebrates as a "standard" in Lamarck's theory of evolution, he believed that characteristics acquired through the evolutionary process involved not only survival, but also progression toward a "higher form", to which humans and vertebrates were closer than invertebrates were. Although goal-directed evolution has been abandoned, the distinction of invertebrates and vertebrates persists to this day, even though the grouping has been noted to be "hardly natural or even very sharp." Another reason cited for this continued distinction is that Lamarck created a precedent through his classifications which is now difficult to escape from. It is also possible that some humans believe that, they themselves being vertebrates, the group deserves more attention than invertebrates. In any event, in the 1968 edition of *Invertebrate Zoology*, it is noted that "division of the Animal Kingdom into vertebrates and invertebrates is artificial and reflects human bias in favor of man's own relatives." The book also points out that the group lumps a vast number of species together, so that no one characteristic describes all invertebrates. In addition, some species included are only remotely related to one another, with some more related to vertebrates than other invertebrates.

In Research

For many centuries, invertebrates have been neglected by biologists, in favor of big vertebrates and "useful" or charismatic species. Invertebrate biology was not a major field of study until the work of Linnaeus and Lamarck in the 18th century. During the 20th century, invertebrate zoology became one of the major fields of natural sciences, with prominent discoveries in the fields of medicine, genetics, palaeontology, and ecology. The study of invertebrates has also benefited law enforcement, as arthropods, and especially insects, were discovered to be a source of information for forensic investigators.

Two of the most commonly studied model organisms nowadays are invertebrates the fruit fly *Drosophila melanogaster* and the nematode *Caenorhabditis elegans*. They have long been the most intensively studied model organisms, and were among the first life-forms to be genetically sequenced. This was facilitated by the severely reduced state of their genomes, but many genes, introns, and linkages have been lost. Analysis of the starlet sea anemone genome has emphasised the importance of sponges, placozoans, and choanoflagellates, also being sequenced, in explaining the arrival of 1500 ancestral genes unique to animals. Invertebrates are also used by scientists in the field of aquatic biomonitoring to evaluate the effects of water pollution and climate change.

Various Invertebrate Animals

Insect

Insects are a class (Insecta) of hexapod invertebrates within the arthropod phylum that have a chitinous exoskeleton, a three-part body (head, thorax and abdomen), three pairs of jointed legs, compound eyes and one pair of antennae. They are the most diverse group of animals on the planet, including more than a million described species and representing more than half of all known living organisms. The number of extant species is estimated at between six and ten million, and potentially represent over 90% of the differing animal life forms on Earth. Insects may be found in nearly all environments, although only a small number of species reside in the oceans, a habitat dominated by another arthropod group, crustaceans.

The life cycles of insects vary but most hatch from eggs. Insect growth is constrained by the inelastic exoskeleton and development involves a series of molts. The immature stages can differ from the adults in structure, habit and habitat, and can include a passive pupal stage in those groups that undergo 4-stage metamorphosis. Insects that undergo 3-stage metamorphosis lack a pupal stage and adults develop through a series of nymphal stages. The higher level relationship of the Hexapoda is unclear. Fossilized insects of enormous size have been found from the Paleozoic Era, including giant dragonflies with wingspans of 55 to 70 cm (22–28 in). The most diverse insect groups appear to have coevolved with flowering plants.

Adult insects typically move about by walking, flying or sometimes swimming. As it allows for rapid yet stable movement, many insects adopt a tripedal gait in which they walk with their legs touching the ground in alternating triangles. Insects are the only invertebrates to have evolved flight. Many insects spend at least part of their lives under water, with larval adaptations that include gills, and some adult insects are aquatic and have adaptations for swimming. Some species, such as water striders, are capable of walking on the surface of water. Insects are mostly solitary, but some, such as certain bees, ants and termites, are social and live in large, well-organized colonies. Some insects, such as earwigs, show maternal care, guarding their eggs and young. Insects can communicate with each other in a variety of ways. Male moths can sense the pheromones of female moths over great distances. Other species communicate with sounds, crickets stridulate, or rub their wings together, to attract a mate and repel other males. Lampyridae in the beetle order communicate with light.

Humans regard certain insects as pests, and attempt to control them using insecticides and a host of other techniques. Some insects damage crops by feeding on sap, leaves or fruits. A few parasitic species are pathogenic. Some insects perform complex ecological roles; blow-flies, for example, help consume carrion but also spread diseases. Insect pollinators are essential to the life cycle of many flowering plant species on which most organisms, including humans, are at least partly dependent; without them, the terrestrial portion of the biosphere (including humans) would be devastated. Many other insects are considered ecologically beneficial as predators and a few provide direct economic benefit. Silkworms and bees have been used extensively by humans for the production of silk and honey, respectively. In some cultures, people eat the larvae or adults of certain insects.

Etymology

The word "insect" comes from the Latin word *insectum*, meaning "with a notched or divided body", or literally "cut into", from the neuter singular perfect passive participle of *insectare*, "to cut into, to cut up", from *in-* "into" and *secare* "to cut"; because insects appear "cut into" three sections. Pliny the Elder introduced the Latin designation as a loan-translation of the Greek word (*éntomos*) or "insect" (as in entomology), which was Aristotle's term for this class of life, also in reference to their "notched" bodies. "Insect" first appears documented in English in 1601 in Holland's translation of Pliny. Translations of Aristotle's term also form the usual word for "insect" in Welsh (trychfil, from *trychu* "to cut" and *mil*, "animal"), Serbo-Croatian (*zareznik*, from *rezati*, "to cut"), Russian (*sekat'*, "to cut"), etc.

Phylogeny and Evolution

The evolutionary relationship of insects to other animal groups remains unclear.

Although traditionally grouped with millipedes and centipedes—possibly on the basis of convergent adaptations to terrestrialisation—evidence has emerged favoring closer evolutionary ties with crustaceans. In the Pancrustacea theory, insects, together with Entognatha, Remipedia, and Cephalocarida, make up a natural clade labeled Miracrustacea.

A report in November 2014 unambiguously places the insects in one clade, with the crustaceans and myriapods, as the nearest sister clades. This study resolved insect phylogeny of all extant insect orders, and provides "a robust phylogenetic backbone tree and reliable time estimates of insect evolution."

Other terrestrial arthropods, such as centipedes, millipedes, scorpions, and spiders, are sometimes confused with insects since their body plans can appear similar, sharing (as do all arthropods) a jointed exoskeleton. However, upon closer examination, their features differ significantly; most noticeably, they do not have the six-legged characteristic of adult insects.

Evolution has produced enormous variety in insects. Pictured are some of the possible shapes of antennae.

The higher-level phylogeny of the arthropods continues to be a matter of debate and research. In 2008, researchers at Tufts University uncovered what they believe is the world's oldest known full-body impression of a primitive flying insect, a 300 million-year-old specimen from the Carboniferous period. The oldest definitive insect fossil is the Devonian *Rhyniognatha hirsti*, from the 396-million-year-old Rhynie chert. It may have superficially resembled a modern-day silverfish insect. This species already possessed dicondylic mandibles (two articulations in the mandible), a feature associated with winged insects, suggesting that wings may already have evolved at this time. Thus, the first insects probably appeared earlier, in the Silurian period.

Four super radiations of insects have occurred, beetles (evolved about 300 million years ago), flies (evolved about 250 million years ago), and moths and wasps (evolved about 150 million years ago). These four groups account for the majority of described species. The flies and moths along with the fleas evolved from the Mecoptera.

The origins of insect flight remain obscure, since the earliest winged insects currently known appear to have been capable fliers. Some extinct insects had an additional pair of winglets attaching to the first segment of the thorax, for a total of three pairs. As of 2009, no evidence suggests the insects were a particularly successful group of animals before they evolved to have wings.

Late Carboniferous and Early Permian insect orders include both extant groups, their stem groups, and a number of Paleozoic groups, now extinct. During this era, some giant dragonfly-like forms reached wingspans of 55 to 70 cm (22 to 28 in), making them far larger than any living insect. This gigantism may have been due to higher atmospheric oxygen levels that allowed increased respiratory efficiency relative to today. The lack of flying vertebrates could have been another factor. Most extinct orders of insects developed during the Permian period that began around 270 million years ago. Many of the early groups became extinct during the Permian-Triassic extinction event, the largest mass extinction in the history of the Earth, around 252 million years ago.

The remarkably successful Hymenoptera appeared as long as 146 million years ago in the Cretaceous period, but achieved their wide diversity more recently in the Cenozoic era, which began 66 million years ago. A number of highly successful insect groups evolved in conjunction with flowering plants, a powerful illustration of coevolution.

Many modern insect genera developed during the Cenozoic. Insects from this period on are often found preserved in amber, often in perfect condition. The body plan, or morphology, of such specimens is thus easily compared with modern species. The study of fossilized insects is called paleoentomology.

Evolutionary Relationships

Insects are prey for a variety of organisms, including terrestrial vertebrates. The earliest vertebrates on land existed 400 million years ago and were large amphibious piscivores. Through gradual evolutionary change, insectivory was the next diet type to evolve.

Insects were among the earliest terrestrial herbivores and acted as major selection agents on plants. Plants evolved chemical defenses against this herbivory and the insects, in turn, evolved mechanisms to deal with plant toxins. Many insects make use of these toxins to protect themselves from their predators. Such insects often advertise their toxicity using warning colors. This successful evolutionary pattern has also been used by mimics. Over time, this has led to complex groups of coevolved species. Conversely, some interactions between plants and insects, like pollination, are beneficial to both organisms. Coevolution has led to the development of very specific mutualisms in such systems.

Taxonomy

Traditional morphology-based or appearance-based systematics have usually given the Hexapoda the rank of superclass, and identified four groups within it; insects (Ectognatha), springtails (Collembola), Protura, and Diplura, the latter three being grouped together as the Entognatha on the basis of internalized mouth parts. Supraordinal relationships have undergone numerous changes with the advent of methods based on evolutionary history and genetic data. A recent theory is that the Hexapoda are polyphyletic (where the last common ancestor was not a member of the group), with the entognath classes having separate evolutionary histories from the Insecta. Many of the traditional appearance-based taxa have been shown to be paraphyletic, so rather than using ranks like subclass, superorder, and infraorder, it has proved better to use monophyletic groupings (in which the last common ancestor is a member of the group). The following represents the best-supported monophyletic groupings for the Insecta.

Insects can be divided into two groups historically treated as subclasses, wingless insects, known as Apterygota, and winged insects, known as Pterygota. The Apterygota consist of the primitively wingless order of the silverfish (Zygentoma). Archaeognatha make up the Monocondylia based on the shape of their mandibles, while Zygentoma and Pterygota are grouped together as Dicondylia. The Zygentoma themselves possibly are not monophyletic, with the family Lepidotrichidae being a sister group to the Dicondylia (Pterygota and the remaining Zygentoma).

Paleoptera and Neoptera are the winged orders of insects differentiated by the presence of hardened body parts called sclerites, and in the Neoptera, muscles that allow their wings to fold flatly over the abdomen. Neoptera can further be divided into incomplete metamorphosis-based (Polyneoptera and Paraneoptera) and complete metamorphosis-based groups. It has proved difficult to clarify the relationships between the orders in Polyneoptera because of constant new findings calling for revision of the taxa. For example, the Paraneoptera have turned out to be more closely related to the Endopterygota than to the rest of the Exopterygota. The recent molecular finding that the traditional louse orders Mallophaga and Anoplura are derived from within Psocoptera has led to the new taxon Psocodea. Phasmatodea and Embiidina have been suggested to form the Eukinolabia. Mantodea, Blattodea, and Isoptera are thought to form a monophyletic group termed Dictyoptera.

The Exopterygota likely are paraphyletic in regard to the Endopterygota. Matters that have incurred controversy include Strepsiptera and Diptera grouped together as Halteria based on a reduction of one of the wing pairs – a position not well-supported in the entomological community. The Neuropterida are often lumped or split on the whims of the taxonomist. Fleas are now thought to be closely related to boreid mecopterans. Many questions remain in the basal relationships amongst endopterygote orders, particularly the Hymenoptera.

The study of the classification or taxonomy of any insect is called systematic entomology. If one works with a more specific order or even a family, the term may also be made specific to that order or family, for example systematic dipterology.

Diversity

Though the true dimensions of species diversity remain uncertain, estimates range from 2.6–7.8 million species with a mean of 5.5 million. This probably represents less than 20% of all species on Earth, and with only about 20,000 new species of all organisms being described each year, most species likely will remain undescribed for many years unless species descriptions increase in rate. About 850,000–1,000,000 of all described species are insects. Of the 24 orders of insects, four dominate in terms of numbers of described species, with at least 670,000 species included in Coleoptera, Diptera, Hymenoptera and Lepidoptera.

Comparison of the estimated number of species in the four most speciose insect orders			
	Described species	**Average description rate (species per year)**	**Publication effort**
Coleoptera	300,000–400,000	2308	0.01
Lepidoptera	180,000	642	0.03
Diptera	90,000–150,000	1048	0.04
Hymenoptera	100,000–150,000	1196	0.02

A 2015 study estimated the number of beetles at 0.9–2.1 million with a mean of 1.5 million.

Morphology and Physiology

External

Insects have segmented bodies supported by exoskeletons, the hard outer covering made mostly of chitin. The segments of the body are organized into three distinctive but interconnected units, or tagmata, a head, a thorax and an abdomen. The head supports a pair of sensory antennae, a pair of compound eyes, and, if present, one to three simple eyes (or ocelli) and three sets of variously modified appendages that form the mouthparts. The thorax has six segmented legs—one pair each for the prothorax, mesothorax and the metathorax segments making up the thorax—and, none, two or four wings. The abdomen consists of eleven segments, though in a few species of insects, these segments may be fused together or reduced in size. The abdomen also contains most of the digestive, respiratory, excretory and reproductive internal structures. Considerable variation and many adaptations in the body parts of insects occur, especially wings, legs, antenna and mouthparts.

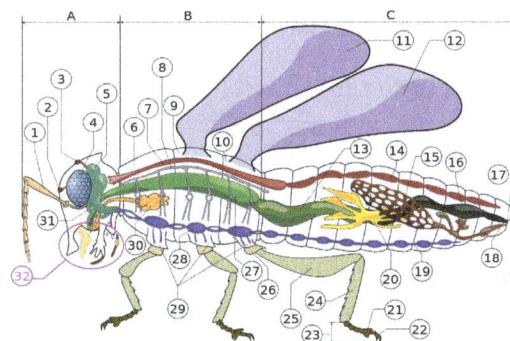

Insect morphology
A- Head **B**- Thorax **C**- Abdomen

1. antenna	9. mesothorax	17. anus	25. femur
2. ocelli (lower)	10. metathorax	18. oviduct	26. trochanter
3. ocelli (upper)	11. forewing	19. nerve chord (abdominal ganglia)	27. fore-gut (crop, gizzard)
4. compound eye	12. hindwing	20. Malpighian tubes	28. thoracic ganglion
5. brain (cerebral ganglia)	13. mid-gut (stomach)	21. tarsal pads	29. coxa
6. prothorax	14. dorsal tube (Heart)	22. claws	30. salivary gland
7. dorsal blood vessel	15. ovary	23. tarsus	31. subesophageal ganglion
8. tracheal tubes (trunk with spiracle)	16. hind-gut (intestine, rectum & anus)	24. tibia	32. mouthparts

Segmentation

The head is enclosed in a hard, heavily sclerotized, unsegmented, exoskeletal head capsule, or epicranium, which contains most of the sensing organs, including the antennae, ocellus or eyes, and the mouthparts. Of all the insect orders, Orthoptera displays the most features found in other insects, including the sutures and sclerites. Here, the vertex, or the apex (dorsal region), is situated between the compound eyes for insects with a hypognathous and opisthognathous head. In prognathous insects, the vertex is not found between the compound eyes, but rather, where the ocelli are normally. This is because the primary axis of the head is rotated 90° to become parallel to the primary axis of the body. In some species, this region is modified and assumes a different name.

The thorax is a tagma composed of three sections, the prothorax, mesothorax and the metathorax. The anterior segment, closest to the head, is the prothorax, with the major features being the first pair of legs and the pronotum. The middle segment is the mesothorax, with the major features being the second pair of legs and the anterior wings. The third and most posterior segment, abutting the abdomen, is the metathorax, which features the third pair of legs and the posterior wings. Each segment is dilineated by an intersegmental suture. Each segment has four basic regions. The dorsal surface is called the tergum (or *notum*) to distinguish it from the abdominal terga. The two lateral regions are called the pleura (singular: pleuron) and the ventral aspect is called the sternum. In turn, the notum of the prothorax is called the pronotum, the notum for the mesothorax is called the mesonotum and the notum for the metathorax is called the metanotum. Continuing with this logic, the mesopleura and metapleura, as well as the mesosternum and metasternum, are used.

The abdomen is the largest tagma of the insect, which typically consists of 11–12 segments and is less strongly sclerotized than the head or thorax. Each segment of the abdomen is represented by a sclerotized tergum and sternum. Terga are separated from each other and from the adjacent sterna or pleura by membranes. Spiracles are located in the pleural area. Variation of this ground plan includes the fusion of terga or terga and sterna to form continuous dorsal or ventral shields or a conical tube. Some insects bear a sclerite in the pleural area called a laterotergite. Ventral sclerites are sometimes called laterosternites. During the embryonic stage of many insects and the postembryonic stage of primitive insects, 11 abdominal segments are present. In modern insects there is a tendency toward reduction in the number of the abdominal segments, but the primitive number of 11 is maintained during embryogenesis. Variation in abdominal segment number is considerable. If the Apterygota are considered to be indicative of the ground plan for pterygotes, confusion reigns; adult Protura have 12 segments, Collembola have 6. The orthopteran family Acrididae has 11 segments, and a fossil specimen of Zoraptera has a 10-segmented abdomen.

Exoskeleton

The insect outer skeleton, the cuticle, is made up of two layers; the epicuticle, which is a thin and waxy water resistant outer layer and contains no chitin, and a lower layer called the procuticle. The procuticle is chitinous and much thicker than the epicuticle and has two layers; an outer layer known as the exocuticle and an inner layer known as the endocuticle. The tough and flexible endocuticle is built from numerous layers of fibrous chitin and proteins, criss-crossing each other in a sandwich pattern, while the exocuticle is rigid and hardened. The exocuticle is greatly reduced in many soft-bodied insects (e.g., caterpillars), especially during their larval stages.

Insects are the only invertebrates to have developed active flight capability, and this has played an important role in their success. Their muscles are able to contract multiple times for each single nerve impulse, allowing the wings to beat faster than would ordinarily be possible. Having their muscles attached to their exoskeletons is more efficient and allows more muscle connections; crustaceans also use the same method, though all spiders use hydraulic pressure to extend their legs, a system inherited from their pre-arthropod ancestors. Unlike insects, though, most aquatic crustaceans are biomineralized with calcium carbonate extracted from the water.

Internal

Nervous System

The nervous system of an insect can be divided into a brain and a ventral nerve cord. The head capsule is made up of six fused segments, each with either a pair of ganglia, or a cluster of nerve cells outside of the brain. The first three pairs of ganglia are fused into the brain, while the three following pairs are fused into a structure of three pairs of ganglia under the insect's esophagus, called the subesophageal ganglion.

The thoracic segments have one ganglion on each side, which are connected into a pair, one pair per segment. This arrangement is also seen in the abdomen but only in the first eight segments. Many species of insects have reduced numbers of ganglia due to fusion or reduction. Some cockroaches have just six ganglia in the abdomen, whereas the wasp *Vespa crabro* has only two in the thorax and three in the abdomen. Some insects, like the house fly *Musca domestica*, have all the body ganglia fused into a single large thoracic ganglion.

At least a few insects have nociceptors, cells that detect and transmit signals responsible for the sensation of pain. This was discovered in 2003 by studying the variation in reactions of larvae of the common fruitfly Drosophila to the touch of a heated probe and an unheated one. The larvae reacted to the touch of the heated probe with a stereotypical rolling behavior that was not exhibited when the larvae were touched by the unheated probe. Although nociception has been demonstrated in insects, there is no consensus that insects feel pain consciously.

Insects are Capable of Learning.

Digestive System

An insect uses its digestive system to extract nutrients and other substances from the food it consumes. Most of this food is ingested in the form of macromolecules and other complex substances like proteins, polysaccharides, fats and nucleic acids. These macromolecules must be broken down by catabolic reactions into smaller molecules like amino acids and simple sugars before being used by cells of the body for energy, growth, or reproduction. This break-down process is known as digestion.

The main structure of an insect's digestive system is a long enclosed tube called the alimentary canal, which runs lengthwise through the body. The alimentary canal directs food unidirectionally from the mouth to the anus. It has three sections, each of which performs a different process of digestion. In addition to the alimentary canal, insects also have paired salivary glands and salivary reservoirs. These structures usually reside in the thorax, adjacent to the foregut.

The salivary glands (element 30 in numbered diagram) in an insect's mouth produce saliva. The salivary ducts lead from the glands to the reservoirs and then forward through the head to an opening called the salivarium, located behind the hypopharynx. By moving its mouthparts (element 32 in numbered diagram) the insect can mix its food with saliva. The mixture of saliva and food then travels through the salivary tubes into the mouth, where it begins to break down. Some insects, like flies, have extra-oral digestion. Insects using extra-oral digestion expel digestive enzymes onto their food to break it down. This strategy allows insects to extract a significant proportion of the available nutrients from the food source. The gut is where almost all of insects' digestion takes place. It can be divided into the foregut, midgut and hindgut.

Foregut

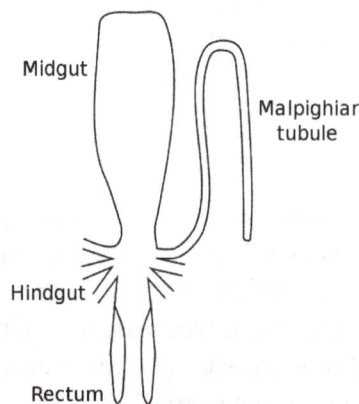

Stylized diagram of insect digestive tract showing malpighian tubule, from an insect of the order Orthoptera

The first section of the alimentary canal is the foregut (element 27 in numbered diagram), or stomodaeum. The foregut is lined with a cuticular lining made of chitin and proteins as protection from tough food. The foregut includes the buccal cavity (mouth), pharynx, esophagus and crop and proventriculus (any part may be highly modified) which both store food and signify when to continue passing onward to the midgut.

Digestion starts in buccal cavity (mouth) as partially chewed food is broken down by saliva from the salivary glands. As the salivary glands produce fluid and carbohydrate-digesting enzymes (mostly amylases), strong muscles in the pharynx pump fluid into the buccal cavity, lubricating the food like the salivarium does, and helping blood feeders, and xylem and phloem feeders.

From there, the pharynx passes food to the esophagus, which could be just a simple tube passing it on to the crop and proventriculus, and then onward to the midgut, as in most insects. Alternately, the foregut may expand into a very enlarged crop and proventriculus, or the crop could just be a diverticulum, or fluid-filled structure, as in some Diptera species.

Bumblebee defecating. Note the contraction of the abdomen to provide internal pressure

Midgut

Once food leaves the crop, it passes to the midgut (element 13 in numbered diagram), also known as the mesenteron, where the majority of digestion takes place. Microscopic projections from the midgut wall, called microvilli, increase the surface area of the wall and allow more nutrients to be absorbed; they tend to be close to the origin of the midgut. In some insects, the role of the microvilli and where they are located may vary. For example, specialized microvilli producing digestive enzymes may more likely be near the end of the midgut, and absorption near the origin or beginning of the midgut.

Hindgut

In the hindgut (element 16 in numbered diagram), or proctodaeum, undigested food particles are joined by uric acid to form fecal pellets. The rectum absorbs 90% of the water in these fecal pellets, and the dry pellet is then eliminated through the anus (element 17), completing the process of digestion. The uric acid is formed using hemolymph waste products diffused from the Malpighian tubules (element 20). It is then emptied directly into the alimentary canal, at the junction between the midgut and hindgut. The number of Malpighian tubules possessed by a given insect varies between species, ranging from only two tubules in some insects to over 100 tubules in others.

Reproductive System

The reproductive system of female insects consist of a pair of ovaries, accessory glands, one or more spermathecae, and ducts connecting these parts. The ovaries are made up of a number of egg tubes, called ovarioles, which vary in size and number by species. The number of eggs that the insect is able to make vary by the number of ovarioles with the rate that eggs can develop being also influenced by ovariole design. Female insects are able make eggs, receive and store sperm, manipulate sperm from different males, and lay eggs. Accessory glands or glandular parts of the oviducts produce a variety of substances for sperm maintenance, transport and fertilization, as well as for protection of eggs. They can produce glue and protective substances for coating eggs or tough coverings for a batch of eggs called oothecae. Spermathecae are tubes or sacs in which sperm can be stored between the time of mating and the time an egg is fertilized.

For males, the reproductive system is the testis, suspended in the body cavity by tracheae and the fat body. Most male insects have a pair of testes, inside of which are sperm tubes or follicles that are enclosed within a membranous sac. The follicles connect to the vas deferens by the vas efferens, and the two tubular vasa deferentia connect to a median ejaculatory duct that leads to the outside. A portion of the vas deferens is often enlarged to form the seminal vesicle, which stores the sperm before they are discharged into the female. The seminal vesicles have glandular linings that secrete nutrients for nourishment and maintenance of the sperm. The ejaculatory duct is derived from an invagination of the epidermal cells during development and, as a result, has a cuticular lining. The terminal portion of the ejaculatory duct may be sclerotized to form the intromittent organ, the aedeagus. The remainder of the male reproductive system is derived from embryonic mesoderm, except for the germ cells, or spermatogonia, which descend from the primordial pole cells very early during embryogenesis.

Respiratory System

The tube-like heart (green) of the mosquito *Anopheles gambiae* extends horizontally across the body, interlinked with the diamond-shaped wing muscles (also green) and surrounded by pericardial cells (red). Blue depicts cell nuclei.

Insect respiration is accomplished without lungs. Instead, the insect respiratory system uses a system of internal tubes and sacs through which gases either diffuse or are actively pumped, delivering oxygen directly to tissues that need it via their trachea (element 8 in numbered diagram). Since oxygen is delivered directly, the circulatory system is not used to carry oxygen, and is therefore greatly reduced. The insect circulatory system has no veins or arteries, and instead consists of little more than a single, perforated dorsal tube which pulses peristaltically. Toward the thorax, the dorsal tube (element 14) divides into chambers and acts like the insect's heart. The opposite end of the dorsal tube is like the aorta of the insect circulating the hemolymph, arthropods' fluid analog of blood, inside the body cavity. Air is taken in through openings on the sides of the abdomen called spiracles.

The respiratory system is an important factor that limits the size of insects. As insects get bigger, this type of oxygen transport gets less efficient and thus the heaviest insect currently weighs less than 100 g. However, with increased atmospheric oxygen levels, as happened in the late Paleozoic, larger insects were possible, such as dragonflies with wingspans of more than two feet.

There are many different patterns of gas exchange demonstrated by different groups of insects. Gas exchange patterns in insects can range from continuous and diffusive ventilation, to discontinuous gas exchange. During continuous gas exchange, oxygen is taken in and carbon dioxide is released in a continuous cycle. In discontinuous gas exchange, however, the insect takes in oxygen while it is active and small amounts of carbon dioxide are released when the insect is at rest. Diffusive ventilation is simply a form of continuous gas exchange that occurs by diffusion rather than physically taking in the oxygen. Some species of insect that are submerged also have adaptations to aid in respiration. As larvae, many insects have gills that can extract oxygen dissolved in water, while others need to rise to the water surface to replenish air supplies which may be held or trapped in special structures.

Circulatory System

The insect circulatory system utilizes hemolymph, a tissue analogous to blood that circulates in the interior of the insect body, while remaining in direct contact with the animal's tissues. It is composed of plasma in which hemocytes are suspended. In addition to hemocytes, the plasma also contains many chemicals. It is also the major tissue type of the open circulatory system of arthropods, characteristic of spiders, crustaceans and insects.

Reproduction and Development

The majority of insects hatch from eggs. The fertilization and development takes place inside the egg, enclosed by a shell (chorion) that consists of maternal tissue. In contrast to eggs of other arthropods, most insect eggs are drought resistant. This is because inside the chorion two additional membranes develop from embryonic tissue, the amnion and the serosa. This serosa secretes a cuticle rich in chitin that protects the embryo against desiccation. In Schizophora however the serosa does not develop, but these flies lay their eggs in damp places, such as rotting matter. Some species of insects, like the cockroach *Blaptica dubia*, as well as juvenile aphids and tsetse flies, are ovoviviparous. The eggs of ovoviviparous animals develop entirely inside the female, and then hatch immediately upon being laid. Some other species, such as those in the genus of cockroaches known as *Diploptera*, are viviparous, and thus gestate inside the mother and are born alive. Some insects, like parasitic wasps, show polyembryony, where a single fertilized egg divides into many and in some cases thousands of separate embryos. Insects may be *univoltine*, *bivoltine* or *multivoltine*, i.e. they may have one, two or many broods (generations) in a year.

A pair of *Simosyrphus grandicornis* hoverflies mating in flight.

Some insects use parthenogenesis, a process in which the female can reproduce and give birth without having the eggs fertilized by a male. Many aphids undergo a form of parthenogenesis, called cyclical parthenogenesis, in which they alternate between one or many generations of asexual and sexual reproduction. In summer, aphids are generally female and parthenogenetic; in the autumn, males may be produced for sexual reproduction. Other insects produced by parthenogenesis are bees, wasps and ants, in which they spawn males. However, overall, most individuals are female, which are produced by fertilization. The males are haploid and the females are diploid. More rarely, some insects display hermaphroditism, in which a given individual has both male and female reproductive organs.

A pair of *grasshoppers* mating.

The different forms of the male (top) and female (bottom) tussock moth *Orgyia recens* is an example of sexual dimorphism in insects.

Insect life-histories show adaptations to withstand cold and dry conditions. Some temperate region insects are capable of activity during winter, while some others migrate to a warmer climate or go into a state of torpor. Still other insects have evolved mechanisms of diapause that allow eggs or pupae to survive these conditions.

Metamorphosis

Metamorphosis in insects is the biological process of development all insects must undergo.

There are two forms of metamorphosis; incomplete metamorphosis and complete metamorphosis.

Incomplete Metamorphosis

Hemimetabolous insects, those with incomplete metamorphosis, change gradually by undergoing a series of molts. An insect molts when it outgrows its exoskeleton, which does not stretch and would otherwise restrict the insect's growth. The molting process begins as the insect's epidermis secretes a new epicuticle inside the old one. After this new epicuticle is secreted, the epidermis releases a mixture of enzymes that digests the endocuticle and thus detaches the old cuticle. When this stage is complete, the insect makes its body swell by taking in a large quantity of water or air, which makes the old cuticle split along predefined weaknesses where the old exocuticle was thinnest.

Immature insects that go through incomplete metamorphosis are called nymphs or in the case of dragonflies and damselflies, also naiads. Nymphs are similar in form to the adult except for the presence of wings, which are not developed until adulthood. With each molt, nymphs grow larger and become more similar in appearance to adult insects.

This southern hawker dragonfly molts its exoskeleton several times during its life as a nymph; shown is the final molt to become a winged adult (eclosion).

Complete Metamorphosis

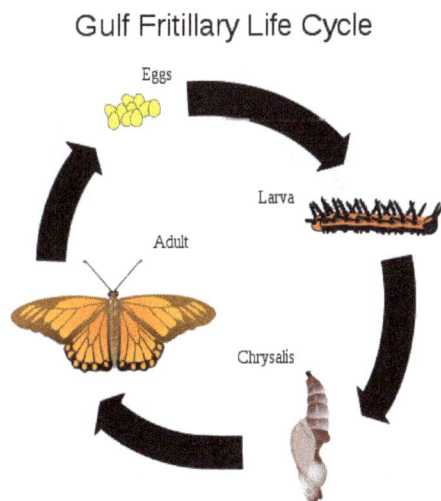

Gulf Fritillary Life Cycle

Eggs

Larva

Adult

Chrysalis

Gulf fritillary life cycle, an example of holometabolism.

Holometabolism, or complete metamorphosis, is where the insect changes in four stages, an egg or embryo, a larva, a pupa and the adult or imago. In these species, an egg hatches to produce a larva, which is generally worm-like in form. This worm-like form can be one of several varieties; eruciform (caterpillar-like), scarabaeiform (grub-like), campodeiform (elongated, flattened and active), elateriform (wireworm-like) or vermiform (maggot-like). The larva grows and eventually becomes a pupa, a stage marked by reduced movement and often sealed within a cocoon. There are three types of pupae; obtect, exarate or coarctate. Obtect pupae are compact, with the legs and other appendages enclosed. Exarate pupae have their legs and other appendages free and extended. Coarctate pupae develop inside the larval skin. Insects undergo considerable change in form during the pupal stage, and emerge as adults. Butterflies are a well-known example of insects that undergo complete metamorphosis, although most insects use this life cycle. Some insects have evolved this system to hypermetamorphosis.

Some of the oldest and most successful insect groups, such Endopterygota, use a system of complete metamorphosis. Complete metamorphosis is unique to a group of certain insect orders including Diptera, Lepidoptera and Hymenoptera. This form of development is exclusive and not seen in any other arthropods.

Senses and Communication

Many insects possess very sensitive and, or specialized organs of perception. Some insects such as bees can perceive ultraviolet wavelengths, or detect polarized light, while the antennae of male moths can detect the pheromones of female moths over distances of many kilometers. The yellow paper wasp (*Polistes versicolor*) is known for its wagging movements as a form of communication within the colony; it can waggle with a frequency of 10.6 ± 2.1 Hz (n=190). These wagging movements can signal the arrival of new material into the nest and aggression between workers can be used to stimulate others to increase foraging expeditions. There is a pronounced tendency for there to be a trade-off between visual acuity and chemical or tactile acuity, such that most insects with well-developed eyes have reduced or simple antennae, and vice versa. There are a variety of different mechanisms by which insects perceive sound, while the patterns are not universal, insects can generally hear sound if they can produce it. Different insect species can have varying hearing, though most insects can hear only a narrow range of frequencies related to the frequency of the sounds they can produce. Mosquitoes have been found to hear up to 2 kHz, and some grasshoppers can hear up to 50 kHz. Certain predatory and parasitic insects can detect the characteristic sounds made by their prey or hosts, respectively. For instance, some nocturnal moths can perceive the ultrasonic emissions of bats, which helps them avoid predation. Insects that feed on blood have special sensory structures that can detect infrared emissions, and use them to home in on their hosts.

Some insects display a rudimentary sense of numbers, such as the solitary wasps that prey upon a single species. The mother wasp lays her eggs in individual cells and provides each egg with a number of live caterpillars on which the young feed when hatched. Some species of wasp always provide five, others twelve, and others as high as twenty-four caterpillars per cell. The number of caterpillars is different among species, but always the same for each sex of larva. The male solitary wasp in the genus *Eumenes* is smaller than the female, so the mother of one species supplies him with only five caterpillars; the larger female receives ten caterpillars in her cell.

Light Production and Vision

A few insects, such as members of the families Poduridae and Onychiuridae (Collembola), Mycetophilidae (Diptera) and the beetle families Lampyridae, Phengodidae, Elateridae and Staphylinidae are bioluminescent. The most familiar group are the fireflies, beetles of the family Lampyridae.

Some species are able to control this light generation to produce flashes. The function varies with some species using them to attract mates, while others use them to lure prey. Cave dwelling larvae of *Arachnocampa* (Mycetophilidae, fungus gnats) glow to lure small flying insects into sticky strands of silk. Some fireflies of the genus *Photuris* mimic the flashing of female *Photinus* species to attract males of that species, which are then captured and devoured. The colors of emitted light vary from dull blue (*Orfelia fultoni*, Mycetophilidae) to the familiar greens and the rare reds (*Phrixothrix tiemanni*, Phengodidae).

Most insects have compound eyes and two antennae.

Most insects, except some species of cave crickets, are able to perceive light and dark. Many species have acute vision capable of detecting minute movements. The eyes may include simple eyes or ocelli as well as compound eyes of varying sizes. Many species are able to detect light in the infrared, ultraviolet and the visible light wavelengths. Color vision has been demonstrated in many species and phylogenetic analysis suggests that UV-green-blue trichromacy existed from at least the Devonian period between 416 and 359 million years ago.

Sound Production and Hearing

Insects were the earliest organisms to produce and sense sounds. Insects make sounds mostly by mechanical action of appendages. In grasshoppers and crickets, this is achieved by stridulation. Cicadas make the loudest sounds among the insects by producing and amplifying sounds with special modifications to their body and musculature. The African cicada *Brevisana brevis* has been measured at 106.7 decibels at a distance of 50 cm (20 in). Some insects, such as the *Helicoverpa zea* moths, hawk moths and Hedylid butterflies, can hear ultrasound and take evasive action when they sense that they have been detected by bats. Some moths produce ultrasonic clicks that were once thought to have a role in jamming bat echolocation. The ultrasonic clicks were subsequently found to be produced mostly by unpalatable moths to warn bats, just as warning colorations are used against predators that hunt by sight. Some otherwise palatable moths have evolved to mimic these calls. More recently, the claim that some moths can jam bat sonar has been revisited. Ultrasonic recording and high-speed infrared videography of bat-moth interactions suggest the palatable tiger moth really does defend against attacking big brown bats using ultrasonic clicks that jam bat sonar.

Very low sounds are also produced in various species of Coleoptera, Hymenoptera, Lepidoptera, Mantodea and Neuroptera. These low sounds are simply the sounds made by the insect's movement. Through microscopic stridulatory structures located on the insect's muscles and joints, the normal sounds of the insect moving are amplified and can be used to warn or communicate with other insects. Most sound-making insects also have tympanal organs that can perceive airborne sounds. Some species in Hemiptera, such as the corixids (water boatmen), are known to communicate via underwater sounds. Most insects are also able to sense vibrations transmitted through surfaces.

Communication using surface-borne vibrational signals is more widespread among insects because of size constraints in producing air-borne sounds. Insects cannot effectively produce low-frequency sounds, and high-frequency sounds tend to disperse more in a dense environment (such as foliage), so insects living in such environments communicate primarily using substrate-borne vibrations. The mechanisms of production of vibrational signals are just as diverse as those for producing sound in insects.

Some species use vibrations for communicating within members of the same species, such as to attract mates as in the songs of the shield bug *Nezara viridula*. Vibrations can also be used to communicate between entirely different species; lycaenid (gossamer-winged butterfly) caterpillars which are myrmecophilous (living in a mutualistic association with ants) communicate with ants in this way. The Madagascar hissing cockroach has the ability to press air through its spiracles to make a hissing noise as a sign of aggression; the death's-head hawkmoth makes a squeaking noise by forcing air out of their pharynx when agitated, which may also reduce aggressive worker honey bee behavior when the two are in close proximity.

Chemical Communication

Chemical communications in animals rely on a variety of aspects including taste and smell. Chemoreception is the physiological response of a sense organ (i.e. taste or smell) to a chemical stimulus where the chemicals act as signals to regulate the state or activity of a cell. A semiochemical is a message-carrying chemical that is meant to attract, repel, and convey information. Types of semiochemicals include pheromones and kairomones. One example is the butterfly *Phengaris arion* which uses chemical signals as a form of mimicry to aid in predation.

In addition to the use of sound for communication, a wide range of insects have evolved chemical means for communication. These chemicals, termed semiochemicals, are often derived from plant metabolites include those meant to attract, repel and provide other kinds of information. Pheromones, a type of semiochemical, are used for attracting mates of the opposite sex, for aggregating conspecific individuals of both sexes, for deterring other individuals from approaching, to mark a trail, and to trigger aggression in nearby individuals. Allomonea benefit their producer by the effect they have upon the receiver. Kairomones benefit their receiver instead of their producer. Synomones benefit the producer and the receiver. While some chemicals are targeted at individuals of the same species, others are used for communication across species. The use of scents is especially well known to have developed in social insects.

Social Behavior

Social insects, such as termites, ants and many bees and wasps, are the most familiar species of eusocial animal. They live together in large well-organized colonies that may be so tightly integrated and genetically similar that the colonies of some species are sometimes considered superorganisms. It is sometimes argued that the various species of honey bee are the only invertebrates (and indeed one of the few non-human groups) to have evolved a system of abstract symbolic communication where a behavior is used to *represent* and convey specific information about something in the environment. In this communication system, called dance language, the angle at which a bee dances represents a direction relative to the sun, and the length of the dance represents the distance to be flown. Though perhaps not as advanced as honey bees, bumblebees also potentially have some social communication behaviors. *Bombus terrestris*, for example, exhibit a faster learning curve for visiting unfamiliar, yet rewarding flowers, when they can see a conspecific foraging on the same species.

A cathedral mound created by termites (Isoptera).

Only insects which live in nests or colonies demonstrate any true capacity for fine-scale spatial orientation or homing. This can allow an insect to return unerringly to a single hole a few millimeters in diameter among thousands of apparently identical holes clustered together, after a trip of up to several kilometers' distance. In a phenomenon known as philopatry, insects that hibernate have shown the ability to recall a specific location up to a year after last viewing the area of interest. A few insects seasonally migrate large distances between different geographic regions (e.g., the overwintering areas of the monarch butterfly).

Care of Young

The eusocial insects build nest, guard eggs, and provide food for offspring full-time. Most insects, however, lead short lives as adults, and rarely interact with one another except to mate or compete for mates. A small number exhibit some form of parental care, where they will at least guard their eggs, and sometimes continue guarding their offspring until adulthood, and possibly even feeding them. Another simple form of parental care is to construct a nest (a burrow or an actual construction, either of which may be simple or complex), store provisions in it, and lay an egg upon those provisions. The adult does not contact the growing offspring, but it nonetheless does provide food. This sort of care is typical for most species of bees and various types of wasps.

Locomotion

Flight

Insects are the only group of invertebrates to have developed flight. The evolution of insect wings has been a subject of debate. Some entomologists suggest that the wings are from paranotal lobes, or extensions from the insect's exoskeleton called the nota, called the *paranotal theory*. Other theories are based on a pleural origin. These theories include suggestions that wings originated from modified gills, spiracular flaps or as from an appendage of the epicoxa. The *epicoxal theory* suggests the insect wings are modified epicoxal exites, a modified appendage at the base of the legs or coxa. In the Carboniferous age, some of the *Meganeura* dragonflies had as much as a 50 cm (20 in) wide wingspan. The appearance of gigantic insects has been found to be consistent with high atmospheric oxygen. The respiratory system of insects constrains their size, however the high oxygen in the atmosphere allowed larger sizes. The largest flying insects today are much smaller and include several moth species such as the Atlas moth and the white witch (*Thysania agrippina*).

White-lined sphinx moth feeding in flight

Insect flight has been a topic of great interest in aerodynamics due partly to the inability of steady-state theories to explain the lift generated by the tiny wings of insects. But insect wings are in motion, with flapping and vibrations, resulting in churning and eddies, and the misconception that physics says "bumblebees can't fly" persisted throughout most of the twentieth century.

Unlike birds, many small insects are swept along by the prevailing winds although many of the larger insects are known to make migrations. Aphids are known to be transported long distances by low-level jet streams. As such, fine line patterns associated with converging winds within weather radar imagery, like the WSR-88D radar network, often represent large groups of insects.

Walking

Many adult insects use six legs for walking and have adopted a tripedal gait. The tripedal gait allows for rapid walking while always having a stable stance and has been studied extensively in cockroaches. The legs are used in alternate triangles touching the ground. For the first step, the middle right leg and the front and rear left legs are in contact with the ground and move the insect forward, while the front and rear right leg and the middle left leg are lifted and moved forward to a new position. When they touch the ground to form a new stable triangle the other legs can be lifted and brought forward in turn and so on. The purest form of the tripedal gait is seen in insects moving at high speeds. However, this type of locomotion is not rigid and insects can adapt a variety of gaits. For example, when moving slowly, turning, or avoiding obstacles, four or more feet may be touching the ground. Insects can also adapt their gait to cope with the loss of one or more limbs.

Cockroaches are among the fastest insect runners and, at full speed, adopt a bipedal run to reach a high velocity in proportion to their body size. As cockroaches move very quickly, they need to be video recorded at several hundred frames per second to reveal their gait. More sedate locomotion is seen in the stick insects or walking sticks (Phasmatodea). A few insects have evolved to walk on the surface of the water, especially members of the Gerridae family, commonly known as water striders. A few species of ocean-skaters in the genus *Halobates* even live on the surface of open oceans, a habitat that has few insect species.

Use in Robotics

Insect walking is of particular interest as an alternative form of locomotion in robots. The study of insects and bipeds has a significant impact on possible robotic methods of transport. This may allow new robots to be designed that can traverse terrain that robots with wheels may be unable to handle.

Swimming

The backswimmer *Notonecta glauca* underwater, showing its paddle-like hindleg adaptation

A large number of insects live either part or the whole of their lives underwater. In many of the more primitive orders of insect, the immature stages are spent in an aquatic environment. Some groups of insects, like certain water beetles, have aquatic adults as well.

Many of these species have adaptations to help in under-water locomotion. Water beetles and water bugs have legs adapted into paddle-like structures. Dragonfly naiads use jet propulsion, forcibly expelling water out of their rectal chamber. Some species like the water striders are capable of walking on the surface of water. They can do this because their claws are not at the tips of the legs as in most insects, but recessed in a special groove further up the leg; this prevents the claws from piercing the water's surface film. Other insects such as the Rove beetle *Stenus* are known to emit pygidial gland secretions that reduce surface tension making it possible for them to move on the surface of water by Marangoni propulsion (also known by the German term *Entspannungsschwimmen*).

Ecology

Insect ecology is the scientific study of how insects, individually or as a community, interact with the surrounding environment or ecosystem. Insects play one of the most important roles in their ecosystems, which includes many roles, such as soil turning and aeration, dung burial, pest control, pollination and wildlife nutrition. An example is the beetles, which are scavengers that feed on dead animals and fallen trees and thereby recycle biological materials into forms found useful by other organisms. These insects, and others, are responsible for much of the process by which topsoil is created.

Defense and Predation

Insects are mostly soft bodied, fragile and almost defenseless compared to other, larger lifeforms. The immature stages are small, move slowly or are immobile, and so all stages are exposed to predation and parasitism. Insects then have a variety of defense strategies to avoid being attacked by predators or parasitoids. These include camouflage, mimicry, toxicity and active defense.

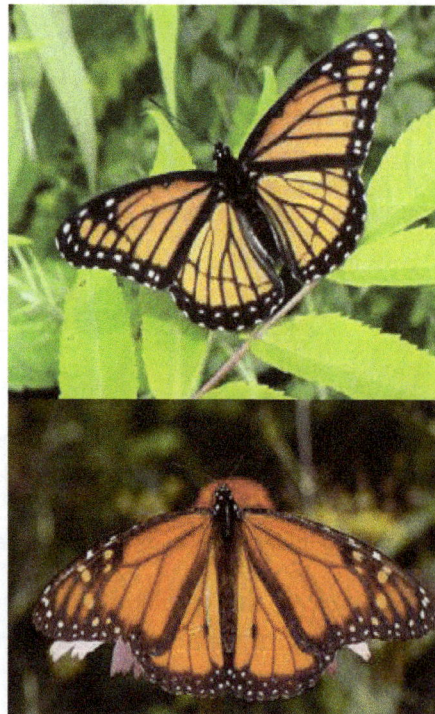

Perhaps one of the most well-known examples of mimicry, the viceroy butterfly (top) appears very similar to the noxious-tasting monarch butterfly (bottom).

Camouflage is an important defense strategy, which involves the use of coloration or shape to blend into the surrounding environment. This sort of protective coloration is common and widespread among beetle families, especially those that feed on wood or vegetation, such as many of the leaf beetles (family Chrysomelidae) or weevils. In some of these species, sculpturing or various colored scales or hairs cause the beetle to resemble bird dung or other inedible objects. Many of those that live in sandy environments blend in with the coloration of the substrate. Most phasmids are known for effectively replicating the forms of sticks and leaves, and the bodies of some species (such as *O. macklotti* and *Palophus centaurus*) are covered in mossy or lichenous outgrowths that supplement their disguise. Some species have the ability to change color as their surroundings shift (*B. scabrinota*, *T. californica*). In a further behavioral adaptation to supplement crypsis, a number of species have been noted to perform a rocking motion where the body is swayed from side to side that is thought to reflect the movement of leaves or twigs swaying in the breeze. Another method by which stick insects avoid predation and resemble twigs is by feigning death (catalepsy), where the insect enters a motionless state that can be maintained for a long period. The nocturnal feeding habits of adults also aids Phasmatodea in remaining concealed from predators.

Another defense that often uses color or shape to deceive potential enemies is mimicry. A number of longhorn beetles (family Cerambycidae) bear a striking resemblance to wasps, which helps them avoid predation even though the beetles are in fact harmless. Batesian and Müllerian mimicry complexes are commonly found in Lepidoptera. Genetic polymorphism and natural selection give rise to otherwise edible species (the mimic) gaining a survival advantage by resembling inedible species (the model). Such a mimicry complex is referred to as *Batesian* and is most commonly known by the mimicry by the limenitidine viceroy butterfly of the inedible danaine monarch. Later research has discovered that the viceroy is, in fact more toxic than the monarch and this resemblance should be considered as a case of Müllerian mimicry. In Müllerian mimicry, inedible species, usually within a taxonomic order, find it advantageous to resemble each other so as to reduce the sampling rate by predators who need to learn about the insects' inedibility. Taxa from the toxic genus *Heliconius* form one of the most well known Müllerian complexes.

Chemical defense is another important defense found amongst species of Coleoptera and Lepidoptera, usually being advertised by bright colors, such as the monarch butterfly. They obtain their toxicity by sequestering the chemicals from the plants they eat into their own tissues. Some Lepidoptera manufacture their own toxins. Predators that eat poisonous butterflies and moths may become sick and vomit violently, learning not to eat those types of species; this is actually the basis of Müllerian mimicry. A predator who has previously eaten a poisonous lepidopteran may avoid other species with similar markings in the future, thus saving many other species as well. Some ground beetles of the Carabidae family can spray chemicals from their abdomen with great accuracy, to repel predators.

Pollination

Pollination is the process by which pollen is transferred in the reproduction of plants, thereby enabling fertilisation and sexual reproduction. Most flowering plants require an animal to do the transportation. While other animals are included as pollinators, the majority of pollination is done by insects. Because insects usually receive benefit for the pollination in the form of energy rich nectar it is a grand example of mutualism. The various flower traits (and combinations thereof) that differentially attract one type of pollinator or another are known as pollination syndromes.

These arose through complex plant-animal adaptations. Pollinators find flowers through bright colorations, including ultraviolet, and attractant pheromones. The study of pollination by insects is known as *anthecology*.

European honey bee carrying pollen in a pollen basket back to the hive

Parasitism

Many insects are parasites of other insects such as the parasitoid wasps. These insects are known as entomophagous parasites. They can be beneficial due to their devastation of pests that can destroy crops and other resources. Many insects have a parasitic relationship with humans such as the mosquito. These insects are known to spread diseases such as malaria and yellow fever and because of such, mosquitoes indirectly cause more deaths of humans than any other animal.

Relationship to Humans

As Pests

Many insects are considered pests by humans. Insects commonly regarded as pests include those that are parasitic (*e.g.* lice, bed bugs), transmit diseases (mosquitoes, flies), damage structures (termites), or destroy agricultural goods (locusts, weevils). Many entomologists are involved in various forms of pest control, as in research for companies to produce insecticides, but increasingly rely on methods of biological pest control, or biocontrol. Biocontrol uses one organism to reduce the population density of another organism — the pest — and is considered a key element of integrated pest management.

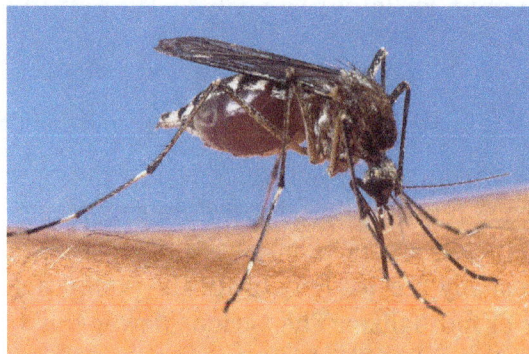

Aedes aegypti, a parasite, is the vector of dengue fever and yellow fever

Despite the large amount of effort focused at controlling insects, human attempts to kill pests with insecticides can backfire. If used carelessly, the poison can kill all kinds of organisms in the area, including insects' natural predators, such as birds, mice and other insectivores. The effects of DDT's use exemplifies how some insecticides can threaten wildlife beyond intended populations of pest insects.

In Beneficial Roles

Because they help flowering plants to cross-pollinate, some insects are critical to agriculture. This European honey bee is gathering nectar while pollen collects on its body.

Although pest insects attract the most attention, many insects are beneficial to the environment and to humans. Some insects, like wasps, bees, butterflies and ants, pollinate flowering plants. Pollination is a mutualistic relationship between plants and insects. As insects gather nectar from different plants of the same species, they also spread pollen from plants on which they have previously fed. This greatly increases plants' ability to cross-pollinate, which maintains and possibly even improves their evolutionary fitness. This ultimately affects humans since ensuring healthy crops is critical to agriculture. As well as pollination ants help with seed distribution of plants. This helps to spread the plants which increases plant diversity. This leads to an overall better environment. A serious environmental problem is the decline of populations of pollinator insects, and a number of species of insects are now cultured primarily for pollination management in order to have sufficient pollinators in the field, orchard or greenhouse at bloom time. Another solution, as shown in Delaware, has been to raise native plants to help support native pollinators like *L. vierecki*. Insects also produce useful substances such as honey, wax, lacquer and silk. Honey bees have been cultured by humans for thousands of years for honey, although contracting for crop pollination is becoming more significant for beekeepers. The silkworm has greatly affected human history, as silk-driven trade established relationships between China and the rest of the world.

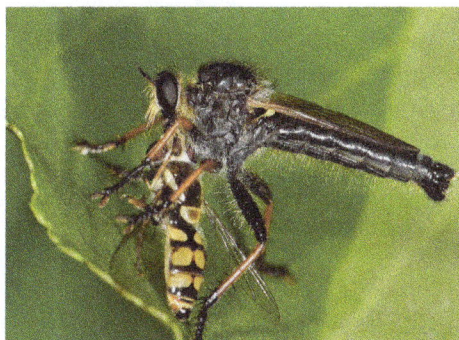

A robberfly with its prey, a hoverfly. Insectivorous relationships such as these help control insect populations.

Insectivorous insects, or insects which feed on other insects, are beneficial to humans because they eat insects that could cause damage to agriculture and human structures. For example, aphids feed on crops and cause problems for farmers, but ladybugs feed on aphids, and can be used as a means to get significantly reduce pest aphid populations. While birds are perhaps more visible predators of insects, insects themselves account for the vast majority of insect consumption. Ants also help control animal populations by consuming small vertebrates. Without predators to keep them in check, insects can undergo almost unstoppable population explosions.

Insects are also used in medicine, for example fly larvae (maggots) were formerly used to treat wounds to prevent or stop gangrene, as they would only consume dead flesh. This treatment is finding modern usage in some hospitals. Recently insects have also gained attention as potential sources of drugs and other medicinal substances. Adult insects, such as crickets and insect larvae of various kinds, are also commonly used as fishing bait.

In Research

Insects play important roles in biological research. For example, because of its small size, short generation time and high fecundity, the common fruit fly *Drosophila melanogaster* is a model organism for studies in the genetics of higher eukaryotes. *D. melanogaster* has been an essential part of studies into principles like genetic linkage, interactions between genes, chromosomal genetics, development, behavior and evolution. Because genetic systems are well conserved among eukaryotes, understanding basic cellular processes like DNA replication or transcription in fruit flies can help to understand those processes in other eukaryotes, including humans. The genome of *D. melanogaster* was sequenced in 2000, reflecting the organism's important role in biological research. It was found that 70% of the fly genome is similar to the human genome, supporting the evolution theory.

As Food

In some cultures, insects, especially deep-fried cicadas, are considered to be delicacies, whereas in other places they form part of the normal diet. Insects have a high protein content for their mass, and some authors suggest their potential as a major source of protein in human nutrition. In most first-world countries, however, entomophagy (the eating of insects), is taboo. Since it is impossible to entirely eliminate pest insects from the human food chain, insects are inadvertently present in many foods, especially grains. Food safety laws in many countries do not prohibit insect parts in food, but rather limit their quantity. According to cultural materialist anthropologist Marvin Harris, the eating of insects is taboo in cultures that have other protein sources such as fish or livestock.

Due to the abundance of insects and a worldwide concern of food shortages, the Food and Agriculture Organisation of the United Nations considers that the world may have to, in the future, regard the prospects of eating insects as a food staple. Insects are noted for their nutrients, having a high content of protein, minerals and fats and are eaten by one-third of the global population.

In Culture

Scarab beetles held religious and cultural symbolism in Old Egypt, Greece and some shamanistic Old World cultures. The ancient Chinese regarded cicadas as symbols of rebirth or immortality. In Mesopotamian literature, the epic poem of Gilgamesh has allusions to Odonata which signify the impossibility of immortality. Amongst the Aborigines of Australia of the Arrernte language groups, honey ants and witchety grubs served as personal clan totems. In the case of the 'San' bush-men of the Kalahari, it is the praying mantis which holds much cultural significance including creation and zen-like patience in waiting.

Mollusca

The molluscs compose the large phylum Mollusca of invertebrate animals. Around 85,000 extant species of molluscs are recognized. Molluscs are the largest marine phylum, comprising about 23% of all the named marine organisms. Numerous molluscs also live in freshwater and terrestrial habitats. They are highly diverse, not just in size and in anatomical structure, but also in behaviour and in habitat. The phylum is typically divided into 9 or 10 taxonomic classes, of which two are entirely extinct. Cephalopod molluscs, such as squid, cuttlefish and octopus, are among the most neurologically advanced of all invertebrates—and either the giant squid or the colossal squid is the largest known invertebrate species. The gastropods (snails and slugs) are by far the most numerous molluscs in terms of classified species, and account for 80% of the total.

Cornu aspersum (formerly *Helix aspersa*) – a common land snail

The three most universal features defining modern molluscs are a mantle with a significant cavity used for breathing and excretion, the presence of a radula (except for bivalves), and the structure of the nervous system. Other than these things, molluscs express great morphological diversity, so many textbooks base their descriptions on a "hypothetical ancestral mollusc" . This has a single, "limpet-like" shell on top, which is made of proteins and chitin reinforced with calcium carbonate, and is secreted by a mantle covering the whole upper surface. The underside of the animal consists of a single muscular "foot". Although molluscs are coelomates, the coelom tends to be small. The main body cavity is a hemocoel through which blood circulates; their circulatory systems are mainly open. The "generalized" mollusc's feeding system consists of a rasping "tongue", the radula, and a complex digestive system in which exuded mucus and microscopic, muscle-powered "hairs" called cilia play various important roles. The generalized mollusc has two paired nerve cords, or three in bivalves. The brain, in species that have one, encircles the esophagus. Most molluscs have eyes, and all have sensors to detect chemicals, vibrations, and touch. The simplest type of molluscan reproductive system relies on external fertilization, but more complex variations occur. All produce eggs, from which may emerge trochophore larvae, more complex veliger larvae, or miniature adults.

Good evidence exists for the appearance of gastropods, cephalopods and bivalves in the Cambrian period 541 to 485.4 million years ago. However, the evolutionary history both of molluscs' emergence from the ancestral Lophotrochozoa and of their diversification into the well-known living and fossil forms are still subjects of vigorous debate among scientists.

Molluscs have been and still are an important food source for anatomically modern humans. However there is a risk of food poisoning from toxins which can accumulate in certain molluscs under specific conditions, and because of this, many countries have regulations to reduce this risk. Molluscs have, for centuries, also been the source of important luxury goods, notably pearls, mother of pearl, Tyrian purple dye, and sea silk. Their shells have also been used as money in some preindustrial societies.

Mollusc species can also represent hazards or pests for human activities. The bite of the blue-ringed octopus is often fatal, and that of *Octopus apollyon* causes inflammation that can last for over a month. Stings from a few species of large tropical cone shells can also kill, but their sophisticated, though easily produced, venoms have become important tools in neurological research. Schistosomiasis (also known as bilharzia, bilharziosis or snail fever) is transmitted to humans via water snail hosts, and affects about 200 million people. Snails and slugs can also be serious agricultural pests, and accidental or deliberate introduction of some snail species into new environments has seriously damaged some ecosystems.

Etymology

The words mollusc and mollusk are both derived from the French *mollusque*, which originated from the Latin *molluscus*, from *mollis*, soft. *Molluscus* was itself an adaptation of Aristotle's (*ta malaká*), "the soft things", which he applied to cuttlefish. The scientific study of molluscs is accordingly called malacology.

The name Molluscoida was formerly used to denote a division of the animal kingdom containing the brachiopods, bryozoans, and tunicates, the members of the three groups having been supposed to somewhat resemble the molluscs. As it is now known these groups have no relation to molluscs, and very little to one another, the name Molluscoida has been abandoned.

Definition

The most universal features of the body structure of molluscs are a mantle with a significant cavity used for breathing and excretion, and the organization of the nervous system. Many have a calcareous shell.

Molluscs have developed such a varied range of body structures, it is difficult to find synapomorphies (defining characteristics) to apply to all modern groups. The most general characteristic of molluscs is they are unsegmented and bilaterally symmetrical. The following are present in all modern molluscs;

The dorsal part of the body wall is a mantle (or pallium) which secretes calcareous spicules, plates or shells. It overlaps the body with enough spare room to form a mantle cavity.

The anus and genitals open into the mantle cavity.

There are two pairs of main nerve cords.

Other characteristics that commonly appear in textbooks have significant exceptions:

Supposed universal Molluscan characteristic	Whether characteristic is found in these classes of Molluscs						
	Aplacophora	Polyplacophora	Monoplacophora	Gastropoda	Cephalopoda	Bivalvia	Scaphopoda
Radula, a rasping "tongue" with chitinous teeth	Absent in 20% of Neomeniomorpha	Yes	Yes	Yes	Yes	No	Internal, cannot extend beyond body
Broad, muscular foot	Reduced or absent	Yes	Yes	Yes	Modified into arms	Yes	Small, only at "front" end
Dorsal concentration of internal organs (visceral mass)	Not obvious	Yes	Yes	Yes	Yes	Yes	Yes
Large digestive ceca	No ceca in some Aplacophora	Yes	Yes	Yes	Yes	Yes	No
Large complex metanephridia ("kidneys")	None	Yes	Yes	Yes	Yes	Yes	Small, simple
One or more valves/ shells	Primitive forms, yes; modern forms, no	Yes	Yes	Snails, yes; slugs, mostly yes (internal vestigial)	Octopuses, no; cuttlefish, nautilus, squid, yes	Yes	Yes
Odontophore	Yes	Yes	Yes	Yes	Yes	No	Yes

Diversity

Estimates of accepted described living species of molluscs vary from 50,000 to a maximum of 120,000 species. In 1969 David Nicol estimated the probable total number of living molluscs at 107,000 of which were about 12,000 fresh-water gastropods and 35,000 terrestrial. The Bivalvia would comprise about 14% of the total and the other five classes less than 2% of the living molluscs. In 2009, Chapman estimated the number of described living species at 85,000. Haszprunar in 2001 estimated about 93,000 named species, which include 23% of all named marine organisms. Molluscs are second only to arthropods in numbers of living animal species—far behind the arthropods' 1,113,000 but well ahead of chordates' 52,000. About 200,000 living species in total are estimated, and 70,000 fossil species, although the total number of mollusc species ever to have existed, whether or not preserved, must be many times greater than the number alive today.

Molluscs have more varied forms than any other animal phylum. They include snails, slugs and other gastropods; clams and other bivalves; squids and other cephalopods; and other lesser-known but similarly distinctive subgroups. The majority of species still live in the oceans, from the seashores to the abyssal zone, but some form a significant part of the freshwater fauna and the terrestrial ecosystems. Molluscs are extremely diverse in tropical and temperate regions, but can be found at all latitudes. About 80% of all known mollusc species are gastropods. Cephalopoda such as squid, cuttlefish, and octopuses are among the neurologically most advanced of all invertebrates. The giant squid, which until recently had not been observed alive in its adult form, is one of the largest invertebrates, but a recently caught specimen of the colossal squid, 10 m (33 ft) long and weighing 500 kg (1,100 lb), may have overtaken it.

About 80% of all known mollusc species are gastropods (snails and slugs), including this cowry (a sea snail).

Freshwater and terrestrial molluscs appear exceptionally vulnerable to extinction. Estimates of the numbers of nonmarine molluscs vary widely, partly because many regions have not been thoroughly surveyed. There is also a shortage of specialists who can identify all the animals in any one area to species. However, in 2004 the IUCN Red List of Threatened Species included nearly 2,000 endangered nonmarine molluscs. For comparison, the great majority of mollusc species are marine, but only 41 of these appeared on the 2004 Red List. About 42% of recorded extinctions since the year 1500 are of molluscs, consisting almost entirely of nonmarine species.

Hypothetical Ancestral Mollusc

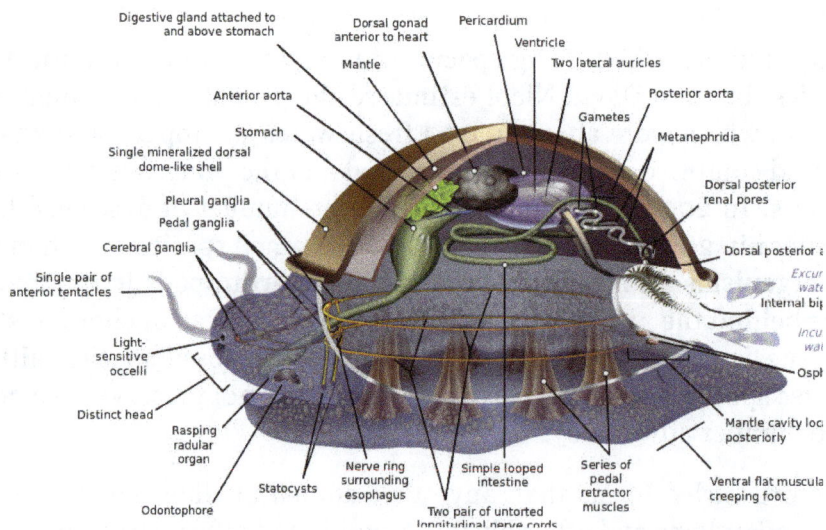

Anatomical diagram of a hypothetical ancestral mollusc

Because of the great range of anatomical diversity among molluscs, many textbooks start the subject of molluscan anatomy by describing what is called an *archi-mollusc*, *hypothetical generalized mollusc*, or *hypothetical ancestral mollusc* (*HAM*) to illustrate the most common features found within the phylum. The depiction is visually rather similar to modern monoplacophorans, and some suggest it may resemble very early molluscs.

The generalized mollusc is bilaterally symmetrical and has a single, "limpet-like" shell on top. The shell is secreted by a mantle covering the upper surface. The underside consists of a single muscular "foot". The visceral mass, or visceropallium, is the soft, nonmuscular metabolic region of the mollusc. It contains the body organs.

Mantle and Mantle Cavity

The mantle cavity, a fold in the mantle, encloses a significant amount of space. It is lined with epidermis, and is exposed, according to habitat, to sea, fresh water or air. The cavity was at the rear in the earliest molluscs, but its position now varies from group to group. The anus, a pair of osphradia (chemical sensors) in the incoming "lane", the hindmost pair of gills and the exit openings of the nephridia ("kidneys") and gonads (reproductive organs) are in the mantle cavity. The whole soft body of bivalves lies within an enlarged mantle cavity.

Shell

The mantle edge secretes a shell (secondarily absent in a number of taxonomic groups, such as the nudibranchs) that consists of mainly chitin and conchiolin (a protein hardened with calcium carbonate), except the outermost layer in almost all cases is all conchiolin. Molluscs never use phosphate to construct their hard parts, with the questionable exception of *Cobcrephora*. While most mollusc shells are composed mainly of aragonite, those gastropods that lay eggs with a hard shell use calcite (sometimes with traces of aragonite) to construct the eggshells.

The shell consists of three layers; the outer layer (the periostracum) made of organic matter, a middle layer made of columnar calcite, and an inner layer consisting of laminated calcite, often nacreous.

Foot

A 50-second video of snails (most likely *Natica chemnitzi* and *Cerithium stercusmuscaram*) feeding on the sea floor in the Gulf of California, Puerto Peñasco, Mexico.

The underside consists of a muscular foot, which has adapted to different purposes in different classes. The foot carries a pair of statocysts, which act as balance sensors. In gastropods, it secretes mucus as a lubricant to aid movement. In forms having only a top shell, such as limpets, the foot acts as a sucker attaching the animal to a hard surface, and the vertical muscles clamp the shell down over it; in other molluscs, the vertical muscles pull the foot and other exposed soft parts into the shell. In bivalves, the foot is adapted for burrowing into the sediment; in cephalopods it is used for jet propulsion, and the tentacles and arms are derived from the foot.

Circulatory System

Molluscs' circulatory systems are mainly open. Although molluscs are coelomates, their coeloms are reduced to fairly small spaces enclosing the heart and gonads. The main body cavity is a hemocoel through which blood and coelomic fluid circulate and which encloses most of the other internal organs. These hemocoelic spaces act as an efficient hydrostatic skeleton. The blood contains the respiratory pigment hemocyanin as an oxygen-carrier. The heart consists of one or more pairs of atria (auricles), which receive oxygenated blood from the gills and pump it to the ventricle, which pumps it into the aorta (main artery), which is fairly short and opens into the hemocoel.

The atria of the heart also function as part of the excretory system by filtering waste products out of the blood and dumping it into the coelom as urine. A pair of nephridia ("little kidneys") to the rear of and connected to the coelom extracts any re-usable materials from the urine and dumps additional waste products into it, and then ejects it via tubes that discharge into the mantle cavity.

Respiration

Most molluscs have only one pair of gills, or even only one gill. Generally, the gills are rather like feathers in shape, although some species have gills with filaments on only one side. They divide the mantle cavity so water enters near the bottom and exits near the top. Their filaments have three kinds of cilia, one of which drives the water current through the mantle cavity, while the other two help to keep the gills clean. If the osphradia detect noxious chemicals or possibly sediment entering the mantle cavity, the gills' cilia may stop beating until the unwelcome intrusions have ceased. Each gill has an incoming blood vessel connected to the hemocoel and an outgoing one to the heart.

Eating, Digestion, and Excretion

Members of the mollusc family use intracellular digestion to function. Most molluscs have muscular mouths with radulae, "tongues", bearing many rows of chitinous teeth, which are replaced from the rear as they wear out. The radula primarily functions to scrape bacteria and algae off rocks, and is associated with the odontophore, a cartilaginous supporting organ. The radula is unique to the molluscs and has no equivalent in any other animal.

Molluscs' mouths also contain glands that secrete slimy mucus, to which the food sticks. Beating cilia (tiny "hairs") drive the mucus towards the stomach, so the mucus forms a long string called a "food string".

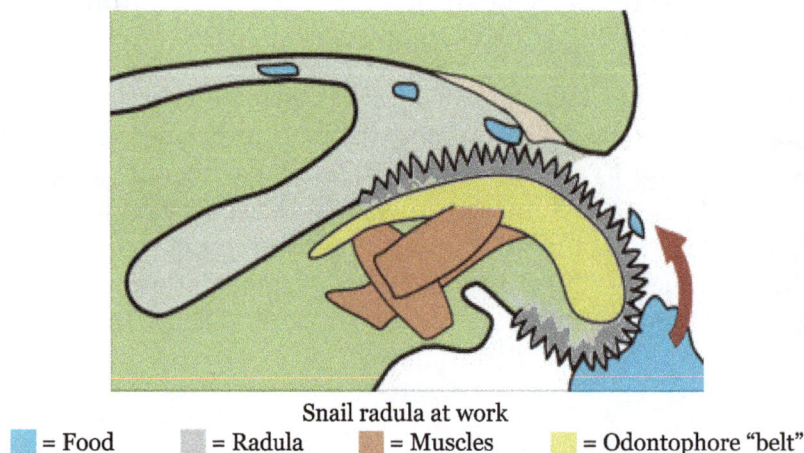

Snail radula at work

■ = Food   ■ = Radula   ■ = Muscles   ■ = Odontophore "belt"

At the tapered rear end of the stomach and projecting slightly into the hindgut is the prostyle, a backward-pointing cone of feces and mucus, which is rotated by further cilia so it acts as a bobbin, winding the mucus string onto itself. Before the mucus string reaches the prostyle, the acidity of the stomach makes the mucus less sticky and frees particles from it.

The particles are sorted by yet another group of cilia, which send the smaller particles, mainly minerals, to the prostyle so eventually they are excreted, while the larger ones, mainly food, are sent to the stomach's cecum (a pouch with no other exit) to be digested. The sorting process is by no means perfect.

Periodically, circular muscles at the hindgut's entrance pinch off and excrete a piece of the prostyle, preventing the prostyle from growing too large. The anus, in the part of the mantle cavity, is swept by the outgoing "lane" of the current created by the gills. Carnivorous molluscs usually have simpler digestive systems.

As the head has largely disappeared in bivalves, the mouth has been equipped with labial palps (two on each side of the mouth) to collect the detritus from its mucus.

Nervous System

The cephalic molluscs have two pairs of main nerve cords organized around a number of paired ganglia, the visceral cords serving the internal organs and the pedal ones serving the foot. Most pairs of corresponding ganglia on both sides of the body are linked by commissures (relatively large bundles of nerves). The ganglia above the gut are the cerebral, the pleural, and the visceral, which are located above the esophagus (gullet). The pedal ganglia, which control the foot, are below the esophagus and their commissure and connectives to the cerebral and pleural ganglia surround the esophagus in a circumesophageal nerve ring or *nerve collar*.

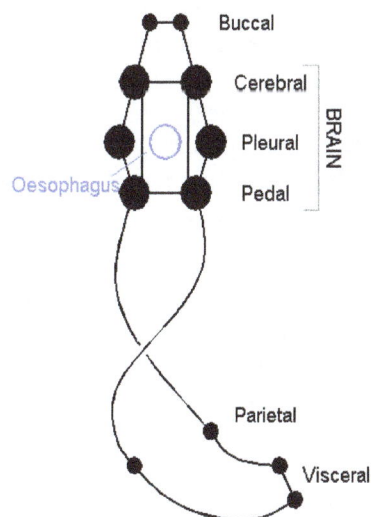

Simplified diagram of the mollusc nervous system

The acephalic molluscs (i.e., bivalves) also have this ring but it is less obvious and less important. The bivalves have only three pairs of ganglia— cerebral, pedal, and visceral— with the visceral as the largest and most important of the three functioning as the principal center of "thinking". Some such as the scallops have eyes around the edges of their shells which connect to a pair of looped nerves and which provide the ability to distinguish between light and shadow.

Reproduction

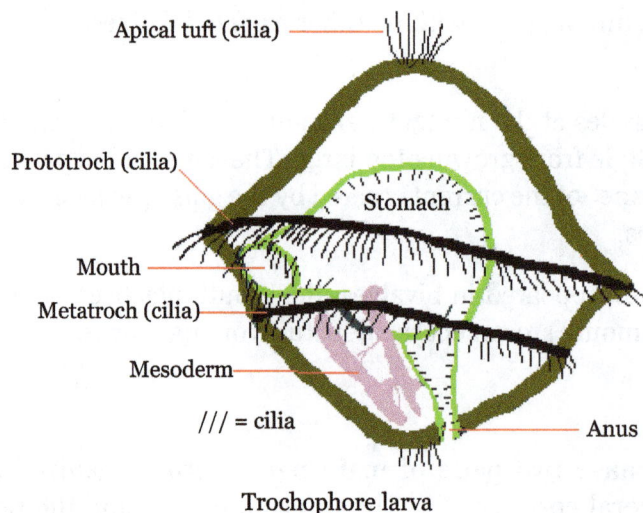

Trochophore larva

The simplest molluscan reproductive system relies on external fertilization, but with more complex variations. All produce eggs, from which may emerge trochophore larvae, more complex veliger larvae, or miniature adults. Two gonads sit next to the coelom, a small cavity that surrounds the heart, into which they shed ova or sperm. The nephridia extract the gametes from the coelom and emit them into the mantle cavity. Molluscs that use such a system remain of one sex all their lives and rely on external fertilization. Some molluscs use internal fertilization and/or are hermaphrodites, functioning as both sexes; both of these methods require more complex reproductive systems.

The most basic molluscan larva is a trochophore, which is planktonic and feeds on floating food particles by using the two bands of cilia around its "equator" to sweep food into the mouth, which uses more cilia to drive them into the stomach, which uses further cilia to expel undigested remains through the anus. New tissue grows in the bands of mesoderm in the interior, so the apical tuft and anus are pushed further apart as the animal grows. The trochophore stage is often succeeded by a veliger stage in which the prototroch, the "equatorial" band of cilia nearest the apical tuft, develops into the velum ("veil"), a pair of cilia-bearing lobes with which the larva swims. Eventually, the larva sinks to the seafloor and metamorphoses into the adult form. While metamorphosis is the usual state in molluscs, the cephalopods differ in exhibiting direct development; the hatchling is a 'miniaturized' form of the adult.

Ecology

Feeding

Most molluscs are herbivorous, grazing on algae or filter feeders. For those grazing, two feeding strategies are predominant. Some feed on microscopic, filamentous algae, often using their radula as a 'rake' to comb up filaments from the sea floor. Others feed on macroscopic 'plants' such as kelp, rasping the plant surface with its radula. To employ this strategy, the plant has to be large enough for the mollusc to 'sit' on, so smaller macroscopic plants are not as often eaten as their larger counterparts. Filter feeders are molluscs that feed by straining suspended matter and food particle from water, typically by passing the water over their gills. Most bivalves are filter feeders.

Cephalopods are primarily predatory, and the radula takes a secondary role to the jaws and tentacles in food acquisition. The monoplacophoran *Neopilina* uses its radula in the usual fashion, but its diet includes protists such as the xenophyophore *Stannophyllum*. Sacoglossan sea-slugs suck the sap from algae, using their one-row radula to pierce the cell walls, whereas dorid nudibranchs and some Vetigastropoda feed on sponges and others feed on hydroids. (An extensive list of molluscs with unusual feeding habits is available in the appendix of *GRAHAM, A. (1955). "Molluscan diets". Journal of Molluscan Studies.*

Classification

Opinions vary about the number of classes of molluscs; for example, the table below shows eight living classes, and two extinct ones. Although they are unlikely to form a clade, some older works combine the Caudofoveata and solenogasters into one class, the Aplacophora. Two of the commonly recognized "classes" are known only from fossils.

Class	Major organisms	Described living species	Distribution
Caudofoveata	worm-like organisms	120	seabed 200–3,000 metres (660–9,840 ft)
Solenogastres	worm-like organisms	200	seabed 200–3,000 metres (660–9,840 ft)
Polyplacophora	chitons	1,000	rocky tidal zone and seabed
Monoplacophora	An ancient lineage of molluscs with cap-like shells	31	seabed 1,800–7,000 metres (5,900–23,000 ft); one species 200 metres (660 ft)
Gastropoda	All the snails and slugs including abalone, limpets, conch, nudibranchs, sea hares, sea butterfly	70,000	marine, freshwater, land
Cephalopoda	squid, octopus, cuttlefish, nautilus, spirula	900	marine
Bivalvia	clams, oysters, scallops, geoducks, mussels	20,000	marine, freshwater
Scaphopoda	tusk shells	500	marine 6–7,000 metres (20–22,966 ft)
Rostroconchia †	fossils; probable ancestors of bivalves	extinct	marine
Helcionelloida †	fossils; snail-like organisms such as *Latouchella*	extinct	marine

Classification into higher taxa for these groups has been and remains problematic. A phylogenetic study suggests the Polyplacophora form a clade with a monophyletic Aplacophora. Additionally, it suggests a sister taxon relationship exists between the Bivalvia and the Gastropoda. Tentaculita may also be in Mollusca.

A genus called *Wiwaxia* probably was in the Mollusca.

Evolution

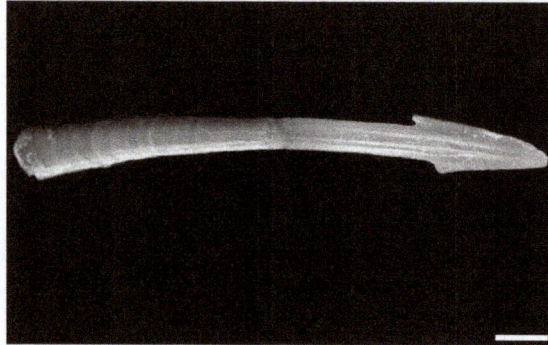

The use of love darts by the land snail *Monachoides vicinus* is a form of sexual selection

Fossil Record

Good evidence exists for the appearance of gastropods, cephalopods and bivalves in the Cambrian period 541 to 485.4 million years ago, or even earlier. The first molluscs appear to have been originated earlier. However, the evolutionary history both of the emergence of molluscs from the ancestral group Lophotrochozoa, and of their diversification into the well-known living and fossil forms, is still vigorously debated.

Debate occurs about whether some Ediacaran and Early Cambrian fossils really are molluscs. *Kimberella*, from about 555 million years ago, has been described by some paleontologists as "mollusc-like", but others are unwilling to go further than "probable bilaterian". There is an even sharper debate about whether *Wiwaxia*, from about 505 million years ago, was a mollusc, and much of this centers on whether its feeding apparatus was a type of radula or more similar to that of some polychaete worms. Nicholas Butterfield, who opposes the idea that *Wiwaxia* was a mollusc, has written that earlier microfossils from 515 to 510 million years ago are fragments of a genuinely mollusc-like radula. This appears to contradict the concept that the ancestral molluscan radula was mineralized.

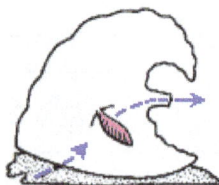

The tiny Helcionellid fossil *Yochelcionella* is thought to be an early mollusc

Spirally coiled shells appear in many gastropods.

However, the Helcionellids, which first appear over 540 million years ago in Early Cambrian rocks from Siberia and China, are thought to be early molluscs with rather snail-like shells. Shelled molluscs therefore predate the earliest trilobites. Although most helcionellid fossils are only a few millimeters long, specimens a few centimeters long have also been found, most with more limpet-like shapes. The tiny specimens have been suggested to be juveniles and the larger ones adults.

Some analyses of helcionellids concluded these were the earliest gastropods. However, other scientists are not convinced these Early Cambrian fossils show clear signs of the torsion that identifies modern gastropods twists the internal organs so the anus lies above the head.

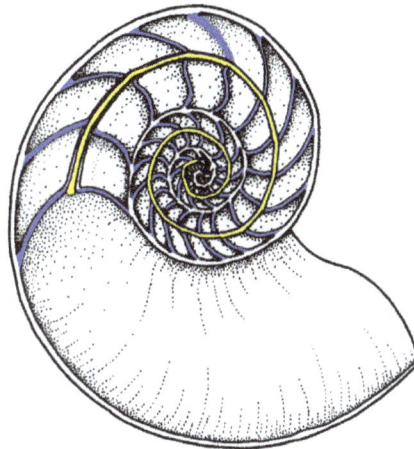

■ = Septa

▨ = Siphuncle

Septa and siphuncle in nautiloid shell

Volborthella, some fossils of which predate 530 million years ago, was long thought to be a cephalopod, but discoveries of more detailed fossils showed its shell was not secreted, but built from grains of the mineral silicon dioxide (silica), and it was not divided into a series of compartments by septa as those of fossil shelled cephalopods and the living *Nautilus* are. *Volborthella*'s classification is uncertain. The Late Cambrian fossil *Plectronoceras* is now thought to be the earliest clearly cephalopod fossil, as its shell had septa and a siphuncle, a strand of tissue that *Nautilus* uses to remove water from compartments it has vacated as it grows, and which is also visible in fossil ammonite shells. However, *Plectronoceras* and other early cephalopods crept along the seafloor instead of swimming, as their shells contained a "ballast" of stony deposits on what is thought to be the underside, and had stripes and blotches on what is thought to be the upper surface. All cephalopods with external shells except the nautiloids became extinct by the end of the Cretaceous period 65 million years ago. However, the shell-less Coleoidea (squid, octopus, cuttlefish) are abundant today.

The Early Cambrian fossils *Fordilla* and *Pojetaia* are regarded as bivalves. "Modern-looking" bivalves appeared in the Ordovician period, 488 to 443 million years ago. One bivalve group, the rudists, became major reef-builders in the Cretaceous, but became extinct in the Cretaceous–Paleogene extinction event. Even so, bivalves remain abundant and diverse.

The Hyolitha are a class of extinct animals with a shell and operculum that may be molluscs. Authors who suggest they deserve their own phylum do not comment on the position of this phylum in the tree of life.

Phylogeny

A possible "family tree" of molluscs (2007). Does not include annelid worms as the analysis concentrated on fossilizable "hard" features.

The phylogeny (evolutionary "family tree") of molluscs is a controversial subject. In addition to the debates about whether *Kimberella* and any of the "halwaxiids" were molluscs or closely related to molluscs, debates arise about the relationships between the classes of living molluscs. In fact, some groups traditionally classified as molluscs may have to be redefined as distinct but related.

Molluscs are generally regarded members of the Lophotrochozoa, a group defined by having trochophore larvae and, in the case of living Lophophorata, a feeding structure called a lophophore. The other members of the Lophotrochozoa are the annelid worms and seven marine phyla. The diagram on the right summarizes a phylogeny presented in 2007.

Because the relationships between the members of the family tree are uncertain, it is difficult to identify the features inherited from the last common ancestor of all molluscs. For example, it is uncertain whether the ancestral mollusc was metameric (composed of repeating units) if it was, that would suggest an origin from an annelid-like worm. Scientists disagree about this; Giribet and colleagues concluded, in 2006, the repetition of gills and of the foot's retractor muscles were later developments, while in 2007, Sigwart concluded the ancestral mollusc was metameric, and it had a foot used for creeping and a "shell" that was mineralized. In one particular branch of the family tree, the shell of conchiferans is thought to have evolved from the spicules (small spines) of aplacophorans; but this is difficult to reconcile with the embryological origins of spicules.

The molluscan shell appears to have originated from a mucus coating, which eventually stiffened into a cuticle. This would have been impermeable and thus forced the development of more sophisticated respiratory apparatus in the form of gills. Eventually, the cuticle would have become mineralized, using the same genetic machinery (engrailed) as most other bilaterian skeletons. The first mollusc shell almost certainly was reinforced with the mineral aragonite.

Morphological analyses tend to recover a conchiferan clade that receives less support from molecular analyses, although these results also lead to unexpected paraphylies, for instance scattering the bivalves throughout all other mollusc groups.

However, an analysis in 2009 using both morphological and molecular phylogenetics comparisons concluded the molluscs are not monophyletic; in particular, Scaphopoda and Bivalvia are both separate, monophyletic lineages unrelated to the remaining molluscan classes; the traditional phylum Mollusca is polyphyletic, and it can only be made monophyletic if scaphopods and bivalves are excluded. A 2010 analysis recovered the traditional conchiferan and aculiferan groups, and showed molluscs were monophyletic, demonstrating that available data for solenogastres was contaminated. Current molecular data are insufficient to constrain the molluscan phylogeny, and since the methods used to determine the confidence in clades are prone to overestimation, it is risky to place too much emphasis even on the areas of which different studies agree. Rather than eliminating unlikely relationships, the latest studies add new permutations of internal molluscan relationships, even bringing the conchiferan hypothesis into question.

Human Interaction

For millennia, molluscs have been a source of food for humans, as well as important luxury goods, notably pearls, mother of pearl, Tyrian purple dye, sea silk, and chemical compounds. Their shells have also been used as a form of currency in some preindustrial societies. A number of species of molluscs can bite or sting humans, and some have become agricultural pests.

Uses by Humans

Molluscs, especially bivalves such as clams and mussels, have been an important food source since at least the advent of anatomically modern humans, and this has often resulted in overfishing. Other commonly eaten molluscs include octopuses and squids, whelks, oysters, and scallops. In 2005, China accounted for 80% of the global mollusc catch, netting almost 11,000,000 tonnes (11,000,000 long tons; 12,000,000 short tons). Within Europe, France remained the industry leader. Some countries regulate importation and handling of molluscs and other seafood, mainly to minimize the poison risk from toxins that can sometimes accumulate in the animals.

Most molluscs with shells can produce pearls, but only the pearls of bivalves and some gastropods, whose shells are lined with nacre, are valuable. The best natural pearls are produced by marine pearl oysters, *Pinctada margaritifera* and *Pinctada mertensi*, which live in the tropical and subtropical waters of the Pacific Ocean. Natural pearls form when a small foreign object gets stuck between the mantle and shell.

Saltwater pearl oyster farm in Seram, Indonesia

The two methods of culturing pearls insert either "seeds" or beads into oysters. The "seed" method uses grains of ground shell from freshwater mussels, and overharvesting for this purpose has endangered several freshwater mussel species in the southeastern United States. The pearl industry is so important in some areas, significant sums of money are spent on monitoring the health of farmed molluscs.

Other luxury and high-status products were made from molluscs. Tyrian purple, made from the ink glands of murex shells, "… fetched its weight in silver" in the fourth century BC, according to Theopompus. The discovery of large numbers of *Murex* shells on Crete suggests the Minoans may have pioneered the extraction of "imperial purple" during the Middle Minoan period in the

20th–18th centuries BC, centuries before the Tyrians. Sea silk is a fine, rare, and valuable fabric produced from the long silky threads (byssus) secreted by several bivalve molluscs, particularly *Pinna nobilis*, to attach themselves to the sea bed. Procopius, writing on the Persian wars *circa* 550 CE, "stated that the five hereditary satraps (governors) of Armenia who received their insignia from the Roman Emperor were given chlamys (or cloaks) made from *lana pinna*. Apparently, only the ruling classes were allowed to wear these chlamys."

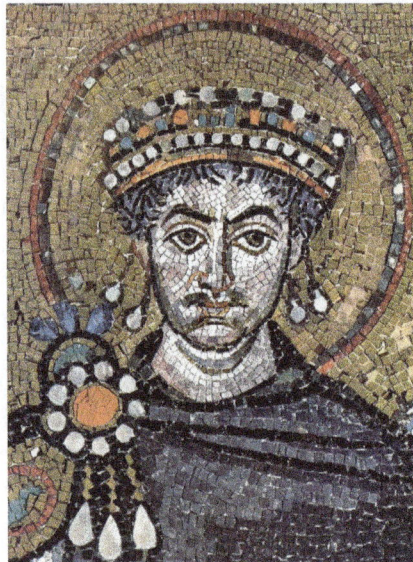

Byzantine Emperor Justinian I clad in Tyrian purple and wearing numerous pearls

Mollusc shells, including those of cowries, were used as a kind of money (shell money) in several preindustrial societies. However, these "currencies" generally differed in important ways from the standardized government-backed and -controlled money familiar to industrial societies. Some shell "currencies" were not used for commercial transactions, but mainly as social status displays at important occasions, such as weddings. When used for commercial transactions, they functioned as commodity money, as a tradable commodity whose value differed from place to place, often as a result of difficulties in transport, and which was vulnerable to incurable inflation if more efficient transport or "goldrush" behavior appeared.

Bioindicators

Bivalve molluscs are used as bioindicators to monitor the health of aquatic environments in both fresh water and the marine environments. Their population status or structure, physiology, behaviour or the level of contamination with elements or compounds can indicate the state of contamination status of the ecosystem. They are particularly useful since they are sessile so that they are representative of the environment where they are sampled or placed.

Harmful to Humans

Stings and Bites

Some molluscs sting or bite, but deaths from mollusc venoms total less than 10% of those from jellyfish stings.

All octopuses are venomous, but only a few species pose a significant threat to humans. Blue-ringed octopuses in the genus *Hapalochlaena*, which live around Australia and New Guinea, bite humans only if severely provoked, but their venom kills 25% of human victims. Another tropical species, *Octopus apollyon*, causes severe inflammation that can last for over a month even if treated correctly, and the bite of *Octopus rubescens* can cause necrosis that lasts longer than one month if untreated, and headaches and weakness persisting for up to a week even if treated.

The blue-ringed octopus's rings are a warning signal; this octopus is alarmed, and its bite can kill.

All species of cone snails are venomous and can sting painfully when handled, although many species are too small to pose much of a risk to humans, and only a few fatalities have been reliably reported. Their venom is a complex mixture of toxins, some fast-acting and others slower but deadlier. The effects of individual cone-shell toxins on victims' nervous systems are so precise as to be useful tools for research in neurology, and the small size of their molecules makes it easy to synthesize them.

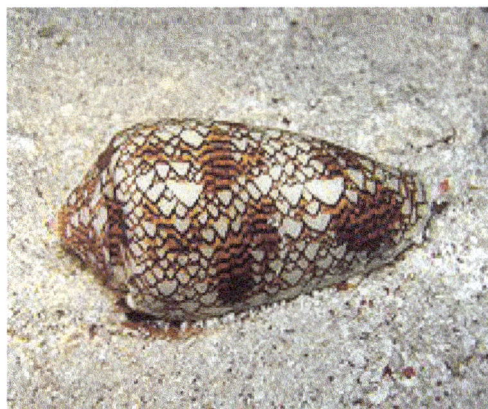

Live cone snails can be dangerous to shell collectors, but are useful to neurology researchers.

Disease Vectors

Schistosomiasis (also known as bilharzia, bilharziosis or snail fever), a disease caused by the fluke worm *Schistosoma*, is "second only to malaria as the most devastating parasitic disease in tropical countries. An estimated 200 million people in 74 countries are infected with the disease – 100 million in Africa alone." The parasite has 13 known species, two of which infect humans. The parasite itself is not a mollusc, but all the species have freshwater snails as intermediate hosts.

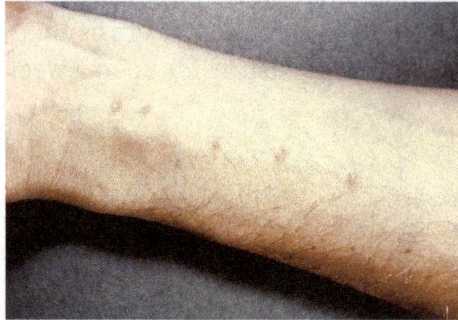

Skin vesicles created by the penetration of *Schistosoma*.

Pests

Some species of molluscs, particularly certain snails and slugs, can be serious crop pests, and when introduced into new environments, can unbalance local ecosystems. One such pest, the giant African snail *Achatina fulica*, has been introduced to many parts of Asia, as well as to many islands in the Indian Ocean and Pacific Ocean. In the 1990s, this species reached the West Indies. Attempts to control it by introducing the predatory snail *Euglandina rosea* proved disastrous, as the predator ignored *Achatina fulica* and went on to extirpate several native snail species, instead.

Marine Invertebrates

Ernst Haeckel's 96th plate, showing some marine invertebrates. Marine invertebrates have a large variety of body plans, which are currently categorised into over 30 phyla.

Marine invertebrates are the invertebrates that live in marine habitats. Invertebrate is a blanket term that includes all animals apart from the vertebrate members of the chordate phylum. Invertebrates lack a vertebral column, and some have evolved a shell or a hard exoskeleton. As on land and in the air, marine invertebrates have a large variety of body plans, and have been categorised into over 30 phyla. They make up most of the macroscopic life in the oceans.

Evolution

Kimberella, an early mollusc important for understanding the Cambrian explosion. Invertebrates are grouped into different phyla (body plans).

The earliest animals were marine invertebrates, that is, vertebrates came later. Animals are multicellular eukaryotes, and are distinguished from plants, algae, and fungi by lacking cell walls. Marine invertebrates are animals that inhabit a marine environment apart from the vertebrate members of the chordate phylum; invertebrates lack a vertebral column. Some have evolved a shell or a hard exoskeleton.

Opabinia an extinct, stem group arthropod that appeared in the Middle Cambrian

The earliest widely accepted animal fossils are the rather modern-looking cnidarians (the group that includes jellyfish, sea anemones and *Hydra*), possibly from around 580 Ma The Ediacara biota, which flourished for the last 40 million years before the start of the Cambrian, were the first animals more than a very few centimetres long. Many were flat and had a "quilted" appearance, and seemed so strange that there was a proposal to classify them as a separate kingdom, Vendozoa. Others, however, have been interpreted as early molluscs (*Kimberella*), echinoderms (*Arkarua*), and arthropods (*Spriggina, Parvancorina*). There is still debate about the classification of these specimens, mainly because the diagnostic features which allow taxonomists to classify more recent organisms, such as similarities to living organisms, are generally absent in the Ediacarans. However, there seems little doubt that *Kimberella* was at least a triploblastic bilaterian animal, in other words, an animal significantly more complex than the cnidarians.

The small shelly fauna are a very mixed collection of fossils found between the Late Ediacaran and Middle Cambrian periods. The earliest, *Cloudina*, shows signs of successful defense against predation and may indicate the start of an evolutionary arms race. Some tiny Early Cambrian shells almost certainly belonged to molluscs, while the owners of some "armor plates," *Halkieria* and *Microdictyon*, were eventually identified when more complete specimens were found in Cambrian lagerstätten that preserved soft-bodied animals.

In the 1970s there was already a debate about whether the emergence of the modern phyla was "explosive" or gradual but hidden by the shortage of Precambrian animal fossils. A re-analysis of fossils from the Burgess Shale lagerstätte increased interest in the issue when it revealed animals, such as *Opabinia*, which did not fit into any known phylum. At the time these were interpreted as evidence that the modern phyla had evolved very rapidly in the Cambrian explosion and that the Burgess Shale's "weird wonders" showed that the Early Cambrian was a uniquely experimental period of animal evolution. Later discoveries of similar animals and the development of new theoretical approaches led to the conclusion that many of the "weird wonders" were evolutionary "aunts" or "cousins" of modern groups—for example that *Opabinia* was a member of the lobopods, a group which includes the ancestors of the arthropods, and that it may have been closely related to the modern tardigrades. Nevertheless, there is still much debate about whether the Cambrian explosion was really explosive and, if so, how and why it happened and why it appears unique in the history of animals.

Classification

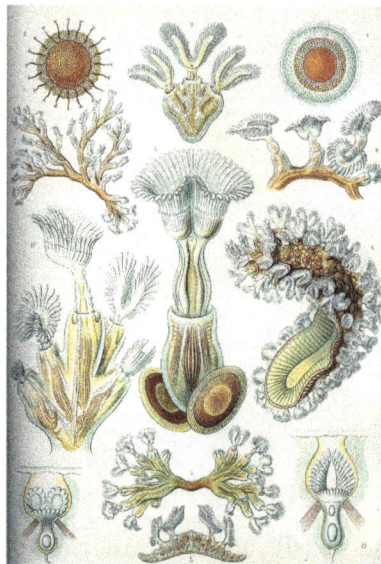

Bryozoa, from Ernst Haeckel's *Kunstformen der Natur*, 1904

Invertebrates are grouped into different phyla. Informally phyla can be thought of as a way of grouping organisms according to their body plan. A body plan refers to a blueprint which describes the shape or morphology of an organism, such as its symmetry, segmentation and the disposition of its appendages. The idea of body plans originated with vertebrates, which were grouped into one phylum. But the vertebrate body plan is only one of many, and invertebrates consist of many phyla or body plans. The history of the discovery of body plans can be seen as a movement from a worldview centred on vertebrates, to seeing the vertebrates as one body plan among many. Among the pioneering zoologists, Linnaeus identified two body plans outside the vertebrates; Cuvier identified three;

and Haeckel had four, as well as the Protista with eight more, for a total of twelve. For comparison, the number of phyla recognised by modern zoologists has risen to 35.

Historically body plans were thought of as having evolved in rapidly during the Cambrian explosion, but a more nuanced understanding of animal evolution suggests a gradual development of body plans throughout the early Palaeozoic and beyond. More generally a phylum can be defined in two ways; as described above, as a group of organisms with a certain degree of morphological or developmental similarity (the phenetic definition), or a group of organisms with a certain degree of evolutionary relatedness (the phylogenetic definition).

As on land and in the air, invertebrates make up a great majority of all macroscopic life, as the vertebrates makes up a subphylum of one of over 30 known animal phyla, making the term almost meaningless for taxonomic purpose. Invertebrate sea life includes the following groups, some of which are phyla:

The 49th plate from Ernst Haeckel's *Kunstformen der Natur*, 1904, showing various sea anemones classified as Actiniae, in the Cnidaria phylum

- Acoela, among the most primitive bilateral animals;

- Annelida, (polychaetes and sea leeches);

- Brachiopoda, marine animals that have hard "valves" (shells) on the upper and lower surfaces ;

- Bryozoa, also known as moss animals or sea mats;

- Chaetognatha, commonly known as arrow worms, are a phylum of predatory marine worms that are a major component of plankton;

- Cnidaria, such as jellyfish, sea anemones, and corals;

- Crustacea, including lobsters, crabs, shrimp, crayfish, barnacles, hermit crabs, mantis shrimps, and copepods;

- Ctenophora, also known as comb jellies, the largest animals that swim by means of cilia;

- Echinodermata, including sea stars, brittle stars, sea urchins, sand dollars, sea cucumbers, crinoids, and sea daisies;

- Echiura, also known as spoon worms;

- Gnathostomulids, slender to thread-like worms, with a transparent body that inhabit sand and mud beneath shallow coastal waters;

- Gastrotricha, often called hairy backs, found mostly interstitially in between sediment particles;

- Hemichordata, includes acorn worms, solitary worm-shaped organisms;

- Kamptozoa, goblet-shaped sessile aquatic animals, with relatively long stalks and a "crown" of solid tentacles, also called Entoprocta;

- Kinorhyncha, segmented, limbless animals, widespread in mud or sand at all depths, also called mud dragons;

- Loricifera, very small to microscopic marine sediment-dwelling animals only discovered in 1983;

- Merostomata; also known as horseshoe crabs;

- Mollusca, including shellfish, squid, octopus, whelks, *Nautilus*, cuttlefish, nudibranchs, scallops, sea snails, Aplacophora, Caudofoveata, Monoplacophora, Polyplacophora, and Scaphopoda;

- Myzostomida, a taxonomic group of small marine worms which are parasitic on crinoids or "sea lilies";

- Nemertinea, also known as "ribbon worms" or "proboscis worms";

- Orthonectida, a small phylum of poorly known parasites of marine invertebrates that are among the simplest of multi-cellular organisms;

- Phoronida, a phylum of marine animals that filter-feed with a lophophore (a "crown" of tentacles), and build upright tubes of chitin to support and protect their soft bodies;

- Placozoa, small, flattened, multicellular animals around 1 millimetre across and the simplest in structure. They have no regular outline, although the lower surface is somewhat concave, and the upper surface is always flattened;

- Porifera (sponges), multicellular organisms that have bodies full of pores and channels allowing water to circulate through them;

- Priapulida, or penis worms, are a phylum of marine worms that live marine mud. They are named for their extensible spiny proboscis, which, in some species, may have a shape like that of a human penis;

- Pycnogonida, also called sea spiders, are unrelated to spiders, or even to arachnids which they resemble;

- Sipunculida, also called peanut worms, is a group containing 144–320 species (estimates vary) of bilaterally symmetrical, unsegmented marine worms;

- Tunicata, also known as sea squirts or sea pork, are filter feeders attached to rocks or similarly suitable surfaces on the ocean floor;

- Some flatworms of the classes Turbellaria and Monogenea;

- Xenoturbella, a genus of bilaterian animals that contains only two marine worm-like species;

- Xiphosura, includes a large number of extinct lineages and only four recent species in the family Limulidae, which include the horseshoe crabs.

"A variety of marine worms": plate from *Das Meer* by M.J. Schleiden (1804–1881)

Arthropods total about 1,113,000 described extant species, molluscs about 85,000 and chordates about 52,000.

Marine Sponges

Sponges have no nervous, digestive or circulatory system

Sponges are animals of the phylum Porifera (Modern Latin for *bearing pores*). They are multi-cellular organisms that have bodies full of pores and channels allowing water to circulate through them, consisting of jelly-like mesohyl sandwiched between two thin layers of cells. They have un-specialized cells that can transform into other types and that often migrate between the main cell layers and the mesohyl in the process. Sponges do not have nervous, digestive or circulatory systems. Instead, most rely on maintaining a constant water flow through their bodies to obtain food and oxygen and to remove wastes.

Sponges are similar to other animals in that they are multicellular, heterotrophic, lack cell walls and produce sperm cells. Unlike other animals, they lack true tissues and organs, and have no body symmetry. The shapes of their bodies are adapted for maximal efficiency of water flow through the central cavity, where it deposits nutrients, and leaves through a hole called the os-culum. Many sponges have internal skeletons of spongin and/or spicules of calcium carbonate or silicon dioxide. All sponges are sessile aquatic animals. Although there are freshwater species, the great majority are marine (salt water) species, ranging from tidal zones to depths exceeding 8,800 m (5.5 mi).

Sponge biodiversity. There are four sponge species in this photo.

While most of the approximately 5,000–10,000 known species feed on bacteria and other food particles in the water, some host photosynthesizing micro-organisms as endosymbionts and these alliances often produce more food and oxygen than they consume. A few species of sponge that live in food-poor environments have become carnivores that prey mainly on small crusta-ceans.

Branching vase sponge

Venus' flower basket at a depth of 2572 meters

Barrel sponge

Stove-pipe sponge

Linnaeus mistakenly identified sponges as plants in the order Algae. For a long time thereafter sponges were assigned to a separate subkingdom, Parazoa (meaning *beside the animals*). They are now classified as a paraphyletic phylum from which the higher animals have evolved.

Marine Cnidarians

Cnidarians are the simplest animals with cells organised into tissues. Yet the starlet sea anemone contains the same genes as those that form the vertebrate head.

Cnidarians (Greek for *nettle*) are distinguished by the presence of stinging cells, specialized cells that they use mainly for capturing prey. Cnidarians include corals, sea anemones, jellyfish and hydrozoans. They form a phylum containing over 10,000 species of animals found exclusively in aquatic (mainly marine) environments. Their bodies consist of mesoglea, a non-living jelly-like substance, sandwiched between two layers of epithelium that are mostly one cell thick. They have two basic body forms; swimming medusae and sessile polyps, both of which are radially symmetrical with mouths surrounded by tentacles that bear cnidocytes. Both forms have a single orifice and body cavity that are used for digestion and respiration.

Fossil cnidarians have been found in rocks formed about 580 million years ago. Fossils of cnidarians that do not build mineralized structures are rare. Scientists currently think cnidarians, ctenophores and bilaterians are more closely related to calcareous sponges than these are to other sponges, and that anthozoans are the evolutionary "aunts" or "sisters" of other cnidarians, and the most closely related to bilaterians.

Cnidarians are the simplest animals in which the cells are organised into tissues. The starlet sea anemone is used as a model organism in research. It is easy to care for in the laboratory and a protocol has been developed which can yield large numbers of embryos on a daily basis. There is a remarkable degree of similarity in the gene sequence conservation and complexity between the sea anemone and vertebrates. In particular, genes concerned in the formation of the head in vertebrates are also present in the anemone.

Sea anemones are common in tidepools

Their tentacles sting and paralyse small fish

Close up of polyps on the surface of a coral, waving their tentacles.

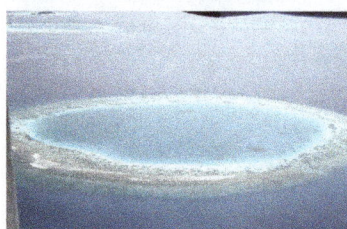
As islands sink below the sea, corals growth can keep pace with the rising water, forming an atoll

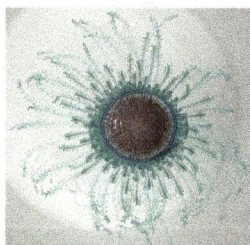
Porpita porpita

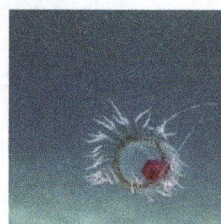
Turritopsis dohrnii, a small, biologically immortal jellyfish transfers its cells back to childhood.

Moon jellyfish, found in coastal waters around the world

Lion's mane jellyfish, largest known jellyfish

Marine Worms

Worms (Old English for *serpent*) typically have long cylindrical tube-like bodies and no limbs. Marine worms vary in size from microscopic to over 1 metre (3.3 ft) in length for some marine polychaete worms (bristle worms) and up to 58 metres (190 ft) for the marine nemertean worm (bootlace worm). Some marine worms occupy a small variety of parasitic niches, living inside the bodies of other animals, while others live more freely in the marine environment or by burrowing underground.

Arrow worms are predatory components of plankton worldwide.

Different groups of marine worms are related only distantly, so they are found in several different phyla such as the Annelida (segmented worms), Chaetognatha (arrow worms), Hemichordata, and Phoronida (horseshoe worms). Many of these worms have specialized tentacles used for exchanging oxygen and carbon dioxide and also may be used for reproduction. Some marine worms are tube worms, such as the giant tube worm which lives in waters near underwater volcanoes and can withstand temperatures up to 90 degrees Celsius.

The bootlace worm can grow to 58 metres (190 ft).

Platyhelminthes (flatworms) form another worm phylum which includes a class Cestoda of parasitic tapeworms. The marine tapeworm *Polygonoporus giganticus*, found in the gut of sperm whales, can grow to over 30 m (100 ft).

Giant tube worms cluster around hydrothermal vents

Nematodes (roundworms) constitute a further worm phylum with tubular digestive systems and an opening at both ends. Over 25,000 nematode species have been described, of which more than half are parasitic. It has been estimated another million remain undescribed. They are ubiquitous in marine, freshwater and terrestrial environments, where they often outnumber other animals in both individual and species counts. They are found in every part of the earth's lithosphere, from the top of mountains to the bottom of oceanic trenches. By count they represent 90% of all animals on the ocean floor. Their numerical dominance, often exceeding a million individuals per square meter and accounting for about 80% of all individual animals on earth, their diversity of life cycles, and their presence at various trophic levels point at an important role in many ecosystems.

Lamellibrachia luymes, a cold seep tubeworm, lives over 250 years.

Bloodworms are typically found on the bottom of shallow marine waters

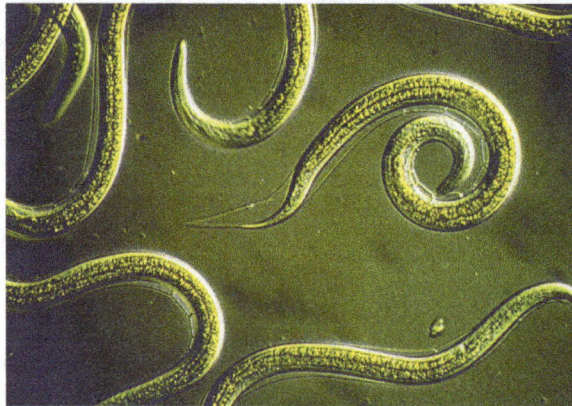
Nematodes are ubiquitous pseudocoelomates which can parasite marine plants and animals.

Echinoderms

Echinoderms (Greek for *spiny skin*) is a phylum which contains only marine invertebrates. The adults are recognizable by their radial symmetry (usually five-point) and include starfish, sea urchins, sand dollars, and sea cucumbers, as well as the sea lilies. Echinoderms are found at every ocean depth, from the intertidal zone to the abyssal zone. The phylum contains about 7000 living species, making it the second-largest grouping of deuterostomes (a superphylum), after the chordates (which include the vertebrates, such as birds, fishes, mammals, and reptiles).

Echinoderms are unique among animals in having bilateral symmetry at the larval stage, but five-fold symmetry (pentamerism, a special type of radial symmetry) as adults.

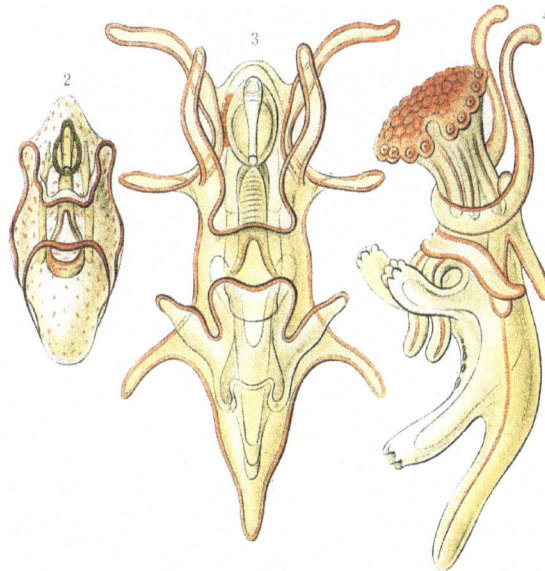

Starfish larvae are bilaterally symmetric, whereas the adults have fivefold symmetry

The echinoderms are important both biologically and geologically. Biologically, there are few other groupings so abundant in the biotic desert of the deep sea, as well as shallower oceans. Most echinoderms are able to regenerate tissue, organs, limbs, and reproduce asexually; in some cases, they can undergo complete regeneration from a single limb. Geologically, the value of echinoderms is in their ossified skeletons, which are major contributors to many limestone formations, and can provide valuable clues as to the geological environment. They were the most used species in regenerative research in the 19th and 20th centuries. Further, it is held by some scientists that the radiation of echinoderms was responsible for the Mesozoic Marine Revolution.

Echinoderm literally means "spiny skin", as this water melon sea urchin illustrates

Sea cucumbers filter feed on plankton and suspended solids

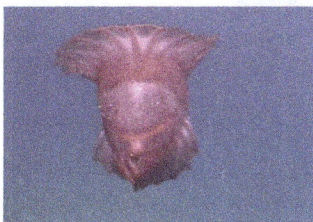

Benthopelagic sea cucumbers can lift off the seafloor and journey as much as 1,000 m (3,300 ft) up the water column

Colorful sea lilies in shallow waters

The ochre sea star was the first keystone predator to be studied. They limit mussels which can overwhelm intertidal communities.

Aside from the hard-to-classify *Arkarua* (a Precambrian animal with echinoderm-like pentamerous radial symmetry), the first definitive members of the phylum appeared near the start of the Cambrian.

Marine Molluscs

Molluscs (Latin for *soft*) form a phylum with about 85,000 extant recognized species. By species count they are the largest marine phylum, comprising about 23% of all the named marine organisms. Molluscs have more varied forms than other invertebrate phylums. They are highly diverse, not just in size and in anatomical structure, but also in behaviour and in habitat. The majority of species still live in the oceans, from the seashores to the abyssal zone, but some form a significant part of the freshwater fauna and the terrestrial ecosystems.

Reconstruction of an ammonite, a highly successful early cephalopod that first appeared in the Devonian (about 400 mya). They became extinct during the same extinction event that killed the land dinosaurs (about 66 mya).

The mollusc phylum is divided into 9 or 10 taxonomic classes, two of which are extinct. These classes include gastropods, bivalves and cephalopods, as well as other lesser-known but distinctive classes. Gastropods with protective shells are referred to as snails, whereas gastropods without protective shells are referred to as slugs. Gastropods are by far the most numerous molluscs in terms of classified species, accounting for 80% of the total. Bivalves include clams, oysters, cockles, mussels, scallops, and numerous other families. There are about 8,000 marine bivalves species (including brackish water and estuarine species), and about 1,200 freshwater species. Cephalopod include octopus, squid and cuttlefish. They are found in all oceans, and neurologically are the most advanced of the invertebrates. About 800 living species of marine cephalopods have been identified, and an estimated 11,000 extinct taxa have been described. There are no fully freshwater cephalopods.

Molluscs have such diverse shapes that many textbooks base their descriptions of molluscan anatomy on a generalized or *hypothetical ancestral mollusc*. This generalized mollusc is unsegmented and bilaterally symmetrical with an underside consisting of a single muscular foot. Beyond that it has three further key features. Firstly, it has a muscular cloak called a mantle covering its viscera and containing a significant cavity used for breathing and excretion. A shell secreted by the mantle covers the upper surface. Secondly (apart from bivalves) it has a rasping tongue called a radula used for feeding. Thirdly, it has a nervous system including a complex digestive system using microscopic, muscle-powered hairs called cilia to exude mucus. The generalized mollusc has two paired nerve cords (three in bivalves). The brain, in species that have one, encircles the esophagus. Most molluscs have eyes and all have sensors detecting chemicals, vibrations, and touch. The simplest type of molluscan reproductive system relies on external fertilization, but more complex variations occur. All produce eggs, from which may emerge trochophore larvae, more complex veliger larvae, or miniature adults. The depiction is rather similar to modern monoplacophorans, and some suggest it may resemble very early molluscs.

Colossal squid, largest of all invertebrates

The nautilus is a living fossil little changed since it evolved 500 million years ago as one of the first cephalopods.

Marine gastropods are sea snails or sea slugs. This nudibranch is a sea slug.

The sea snail *Syrinx aruanus* has the largest shell of any living gastropod

Molluscs usually have eyes. Bordering the edge of the mantle of a scallop, a bivalve mollusc, can be over 100 simple eyes.

Common mussel, another bivalve

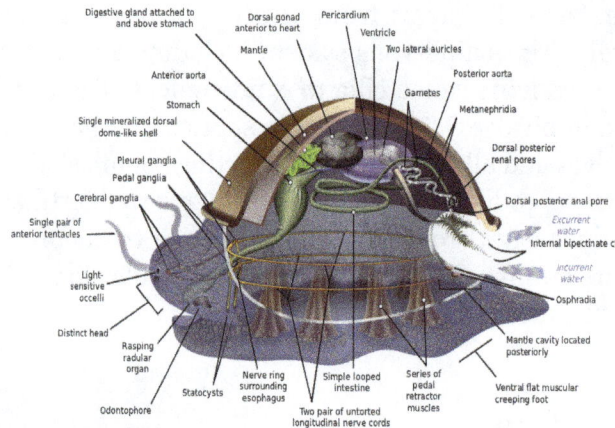

Generalized or *hypothetical ancestral mollusc*

Good evidence exists for the appearance of marine gastropods, cephalopods and bivalves in the Cambrian period 541 to 485.4 million years ago. However, the evolutionary history both of molluscs' emergence from the ancestral Lophotrochozoa and of their diversification into the well-known living and fossil forms are still subjects of vigorous debate among scientists.

Marine Arthropods

Arthropods (Greek for *jointed feet*) have an exoskeleton (external skeleton), a segmented body, and jointed appendages (paired appendages). They form a phylum which includes insects, arachnids, myriapods, and crustaceans. Arthropods are characterized by their jointed limbs and cuticle made of chitin, often mineralised with calcium carbonate. The arthropod body plan consists of segments, each with a pair of appendages. The rigid cuticle inhibits growth, so arthropods replace it periodically by moulting. Their versatility has enabled them to become the most species-rich members of all ecological guilds in most environments.

Marine arthropods range in size from the microscopic crustacean *Stygotantulus* to the Japanese spider crab. Arthropods' primary internal cavity is a hemocoel, which accommodates their internal organs, and through which their haemolymph - analogue of blood - circulates; they have open circulatory systems. Like their exteriors, the internal organs of arthropods are generally built of repeated segments. Their nervous system is "ladder-like", with paired ventral nerve cords running through all segments and forming paired ganglia in each segment. Their heads are formed by fusion of varying numbers of segments, and their brains are formed by fusion of the ganglia of these segments and encircle the esophagus. The respiratory and excretory systems of arthropods vary, depending as much on their environment as on the subphylum to which they belong.

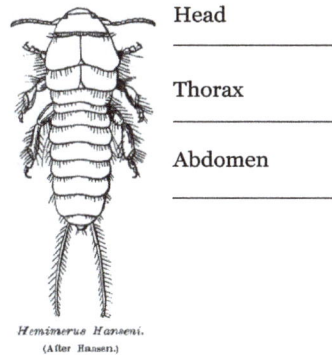

Head

Thorax

Abdomen

Hemimerus Hanseni.
(After Hansen.)

Segmentation and tagmata of an arthropod

Their vision relies on various combinations of compound eyes and pigment-pit ocelli; in most species the ocelli can only detect the direction from which light is coming, and the compound eyes are the main source of information, but the main eyes of spiders are ocelli that can form images and, in a few cases, can swivel to track prey. Arthropods also have a wide range of chemical and mechanical sensors, mostly based on modifications of the many setae (bristles) that project through their cuticles. Arthropods' methods of reproduction and development are diverse; all terrestrial species use internal fertilization, but this is often by indirect transfer of the sperm via an appendage or the ground, rather than by direct injection. Marine species all lay eggs and use either internal or external fertilization. Arthropod hatchlings vary from miniature adults to grubs that lack jointed limbs and eventually undergo a total metamorphosis to produce the adult form.

Trilobites, now extinct, roamed oceans for 270 million years.

Horseshoe crab, a living fossil arthropod from 450 million years ago

Crustaceans

Many crustaceans are very small, like this tiny amphipod, and make up a significant part of the ocean's zooplankton

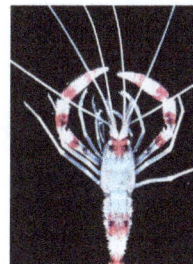

The banded cleaner shrimp is a crustacean common in the tropics.

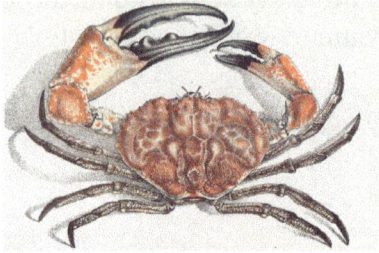

The Tasmanian giant crab is long-lived and slow-grow-ing, making it vulnerable to overfishing.

The Japanese spider crab has the longest leg span of any arthropod.

The evolutionary ancestry of arthropods dates back to the Cambrian period. The group is generally regarded as monophyletic, and many analyses support the placement of arthropods with cycloneura-lians (or their constituent clades) in a superphylum Ecdysozoa. Overall however, the basal relation-ships of Metazoa are not yet well resolved. Likewise, the relationships between various arthropod groups are still actively debated.

Other phyla

- Tardigrade, Lobopodia, (Onychophora)

- Non-craniate (non-vertebrate) chordates: Cephalochordate, Tunicata and *Haikouella*. These invertebrates are close relatives of the vertebrates.

- Non-craniate chordates are close relatives of vertebrates

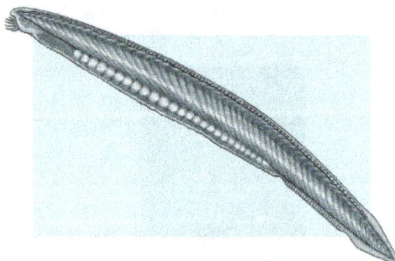

The lancelet, a small translucent fish-like Cephalochor-date, is the closest living invertebrate relative of the vertebrates.

Fluorescent-colored sea squirts, *Rhopalaea crassa*. Tunicates may provide clues to vertebrate (and therefore human) ancestry.

Salp chain

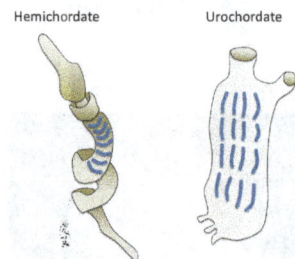

Gill slits in an acorn worm (left) and tunicate (right)

Minerals from Sea Water

There are a number of marine invertebrates that use minerals that are present in the sea in such minute quantities that they were undetectable until the advent of spectroscopy. Vanadium is concentrated by some tunicates for use in their blood cells to a level ten million times that of the surrounding seawater. Other tunicates similarly concentrate niobium and tantalum. Lobsters use copper in their respiratory pigment hemocyanin, despite the proportion of this metal in seawater being minute. Although these elements are present in vast quantities in the ocean, their extraction by man is not economic.

Coral

Corals are marine invertebrates in the class Anthozoa of phylum Cnidaria. They typically live in compact colonies of many identical individual polyps. The group includes the important reef builders that inhabit tropical oceans and secrete calcium carbonate to form a hard skeleton.

A coral "group" is a colony of myriad genetically identical polyps. Each polyp is a sac-like animal typically only a few millimeters in diameter and a few centimeters in length. A set of tentacles surround a central mouth opening. An exoskeleton is excreted near the base. Over many generations, the colony thus creates a large skeleton that is characteristic of the species. Individual heads grow by asexual reproduction of polyps. Corals also breed sexually by spawning; polyps of the same species release gametes simultaneously over a period of one to several nights around a full moon.

Although some corals can catch small fish and plankton using stinging cells on their tentacles, most corals obtain the majority of their energy and nutrients from photosynthetic unicellular dinoflagellates in the genus *Symbiodinium* that live within their tissues. These are commonly known as zooxanthellae and the corals that contain them are zooxanthellate corals. Such corals require sunlight and grow in clear, shallow water, typically at depths shallower than 60 metres (200 ft). Corals are major contributors to the physical structure of the coral reefs that develop in tropical and subtropical waters, such as the enormous Great Barrier Reef off the coast of Queensland, Australia.

Other corals do not rely on zooxanthellae and can live in much deeper water, with the cold-water genus *Lophelia* surviving as deep as 3,000 metres (9,800 ft). Some have been found on the Darwin Mounds, north-west of Cape Wrath, Scotland. Corals have also been found as far north as off the coast of Washington State and the Aleutian Islands.

Taxonomy

In his *Scala Naturae*, Aristotle classified corals as "zoophyta" ("plant-animals"), animals that had characteristics of plants and were therefore hypothetically in between animals and plants. The Persian polymath Al-Biruni (d. 1048) classified sponges and corals as animals, arguing that they respond to touch. Nevertheless, people believed corals to be plants until the eighteenth century, when William Herschel used a microscope to establish that coral had the characteristic thin cell membranes of an animal.

The phylogeny of Anthozoans is not clearly understood and a number of different models have been proposed. Within the Hexacorallia, the sea anemones, coral anemones and stony corals may constitute a monophyletic grouping united by their eight-fold symmetry and cnidocyte trait. The Octocorallia appears to be monophyletic, and primitive members of this group may have been stolonate. The cladogram presented here comes from a 2014 study by Stampar *et al.* which was based on the divergence of mitochondrial DNA within the group and on nuclear markers.

Corals are classified in the class Anthozoa of the phylum Cnidaria. They are divided into three subclasses, Hexacorallia, Octocorallia, and Ceriantharia. The Hexacorallia include the stony corals, the sea anemones and the zoanthids. These groups have polyps that generally have 6-fold symmetry. The Octocorallia include blue coral, soft corals, sea pens, and gorgonians (sea fans and sea whips). These groups have polyps with 8-fold symmetry, each polyp having eight tentacles and eight mesenteries. Ceriantharia are the tube-dwelling anemones.

Fire corals are not true corals, being in the order Anthomedusa (sometimes known as Anthoathecata) of the class Hydrozoa.

Anatomy

Corals are sessile animals in the class Anthozoa and differ from most other cnidarians in not having a medusa stage in their life cycle. The body unit of the animal is a polyp. Most corals are colonial, the initial polyp budding to produce another and the colony gradually developing from this small start. In stony corals, also known as hard corals, the polyps produce a skeleton composed of calcium carbonate to strengthen and protect the organism. This is deposited by the polyps and by the coenosarc, the living tissue that connects them. The polyps sit in cup-shaped depressions in the skeleton known as corallites. Colonies of stony coral are very variable in appearance; a single species may adopt an encrusting, plate-like, bushy, columnar or massive solid structure, the various forms often being linked to different types of habitat, with variations in light level and water movement being significant.

In soft corals, there is no stony skeleton but the tissues are often toughened by the presence of tiny skeletal elements known as sclerites, which are made from calcium carbonate. Soft corals are very variable in form and most are colonial. A few soft corals are stolonate, but the polyps of most are connected by sheets of coenosarc. In some species this is thick and the polyps are deeply embedded. Some soft corals are encrusting or form lobes. Others are tree-like or whip-like and have a central axial skeleton embedded in the tissue matrix. This is composed either of a fibrous protein called gorgonin or of a calcified material. In both stony and soft corals, the polyps can be retracted, with stony corals relying on their hard skeleton and cnidocytes for defence against predators, with soft corals generally relying on chemical defences in the form of toxic substances present in the tissues known as terpenoids.

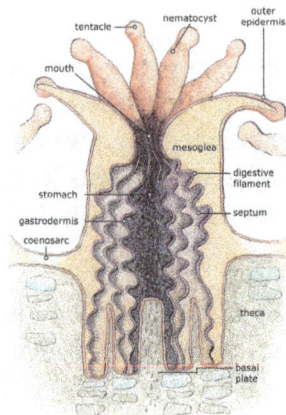

Anatomy of a stony coral polyp

Montastraea cavernosa polyps with tentacles extended

The polyps of stony corals have six-fold symmetry while those of soft corals have eight. The mouth of each polyp is surrounded by a ring of tentacles. In stony corals these are cylindrical and taper to a point, but in soft corals they are pinnate with side branches known as pinnules. In some tropical species these are reduced to mere stubs and in some they are fused to give a paddle-like appearance. In most corals, the tentacles are retracted by day and spread out at night to catch plankton and other small organisms. Shallow water species of both stony and soft corals can be zooxanthellate, the corals supplementing their plankton diet with the products of photosynthesis produced by these symbionts. The polyps interconnect by a complex and well-developed system of gastrovascular canals, allowing significant sharing of nutrients and symbionts.

Ecology

Feeding

Polyps feed on a variety of small organisms, from microscopic zooplankton to small fish. The polyp's tentacles immobilize or kill prey using their nematocysts. These cells carry venom which they rapidly release in response to contact with another organism. A dormant nematocyst discharges in response to nearby prey touching the trigger (cnidocil). A flap (operculum) opens and its stinging apparatus fires the barb into the prey. The venom is injected through the hollow filament to immobilise the prey; the tentacles then manoeuvre the prey to the mouth.

The tentacles then contract to bring the prey into the stomach. Once the prey is digested, the stomach reopens, allowing the elimination of waste products and the beginning of the next hunting cycle. They can scavenge drifting organic molecules and dissolved organic molecules.

Intracellular Symbionts

Many corals, as well as other cnidarian groups such as *Aiptasia* (a sea anemone) form a symbiotic relationship with a class of dinoflagellate algae, zooxanthellae of the genus *Symbiodinium.*:=*Aiptasia*, a familiar pest among coral reef aquarium hobbyists, serves as a valuable model organism in the study of cnidarian-algal symbiosis. Typically, each polyp harbors one species of algae. Via photosynthesis, these provide energy for the coral, and aid in calcification. As much as 30% of the tissue of a polyp may be algal material.

The algae benefit from a safe place to live and consume the polyp's carbon dioxide and nitrogenous waste. Due to the strain the algae can put on the polyp, stress on the coral often drives them to eject the algae. Mass ejections are known as coral bleaching, because the algae contribute to coral's brown coloration; other colors, however, are due to host coral pigments, such as green fluorescent proteins (GFPs). Ejection increases the polyp's chance of surviving short-term stress—they can regain algae, possibly of a different species at a later time. If the stressful conditions persist, the polyp eventually dies.

Reproduction

Corals can be both gonochoristic (unisexual) and hermaphroditic, each of which can reproduce sexually and asexually. Reproduction also allows coral to settle in new areas.

Sexual

Corals predominantly reproduce sexually. About 25% of hermatypic corals (stony corals) form single sex (gonochoristic) colonies, while the rest are hermaphroditic.

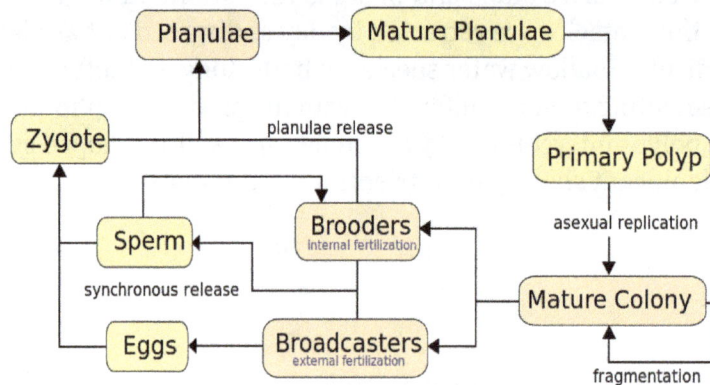

Life cycles of broadcasters and brooders

Broadcasters

About 75% of all hermatypic corals "broadcast spawn" by releasing gametes—eggs and sperm—into the water to spread offspring. The gametes fuse during fertilization to form a microscopic larva called a planula, typically pink and elliptical in shape. A typical coral colony forms several thousand larvae per year to overcome the odds against formation of a new colony.

A male great star coral, *Montastraea cavernosa*, releasing sperm into the water.

Synchronous spawning is very typical on the coral reef, and often, even when multiple species are present, all corals spawn on the same night. This synchrony is essential so male and female gametes can meet. Corals rely on environmental cues, varying from species to species, to determine the proper time to release gametes into the water. The cues involve temperature change, lunar cycle, day length, and possibly chemical signalling. Synchronous spawning may form hybrids and is perhaps involved in coral speciation. The immediate cue is most often sunset, which cues the release. The spawning event can be visually dramatic, clouding the usually clear water with gametes.

Brooders

Brooding species are most often ahermatypic (not reef-building) in areas of high current or wave action. Brooders release only sperm, which is negatively buoyant, sinking on to the waiting egg carriers who harbor unfertilized eggs for weeks. Synchronous spawning events sometimes occurs even with these species. After fertilization, the corals release planula that are ready to settle.

Planulae

Planulae exhibit positive phototaxis, swimming towards light to reach surface waters, where they drift and grow before descending to seek a hard surface to which they can attach and begin a new colony. They also exhibit positive sonotaxis, moving towards sounds that emanate from the reef and away from open water. High failure rates afflict many stages of this process, and even though millions of gametes are released by each colony, few new colonies form. The time from spawning to settling is usually two to three days, but can be up to two months. The larva grows into a polyp and eventually becomes a coral head by asexual budding and growth.

Asexual

Within a coral head, the genetically identical polyps reproduce asexually, either by budding (gemmation) or by dividing, whether longitudinally or transversely.

Basal plates (calices) of *Orbicella annularis* showing multiplication by budding (small central plate) and division (large double plate)

Budding involves splitting a smaller polyp from an adult. As the new polyp grows, it forms its body parts. The distance between the new and adult polyps grows, and with it, the coenosarc (the common body of the colony). Budding can be intratentacular, from its oral discs, producing same-sized polyps within the ring of tentacles, or extratentacular, from its base, producing a smaller polyp.

Tabulate coral *Aulopora* (Devonian) showing initial budding

Division forms two polyps that each become as large as the original. Longitudinal division begins when a polyp broadens and then divides its coelenteron (body), effectively splitting along its length. The mouth divides and new tentacles form. The two polyps thus created then generate their missing body parts and exoskeleton. Transversal division occurs when polyps and the exoskeleton divide transversally into two parts. This means one has the basal disc (bottom) and the other has the oral disc (top); the new polyps must separately generate the missing pieces.

Asexual reproduction offers the benefits of high reproductive rate, delaying senescence, and replacement of dead modules, as well as geographical distribution.

Colony Division

Whole colonies can reproduce asexually, forming two colonies with the same genotype. The possible mechanisms include fission, bailout and fragmentation. Fission occurs in some corals, especially among the family Fungiidae, where the colony splits into two or more colonies during early developmental stages. Bailout occurs when a single polyp abandons the colony and settles on a different substrate to create a new colony. Fragmentation involves individuals broken from the colony during storms or other disruptions. The separated individuals can start new colonies.

Reefs

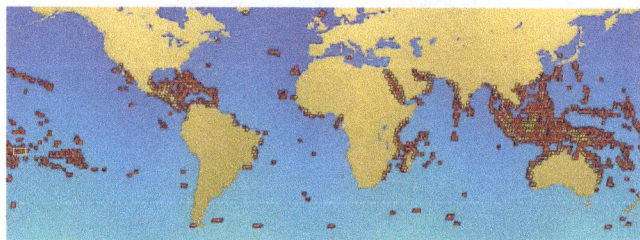

Locations of coral reefs around the world

Many corals in the order Scleractinia are hermatypic, meaning that they are involved in building reefs. Most such corals obtain some of their energy from zooxanthellae in the genus *Symbiodinium*. These are symbiotic photosynthetic dinoflagellates which require sunlight; reef-forming corals are therefore found mainly in shallow water. They secrete calcium carbonate to form hard skeletons that become the framework of the reef. However, not all reef-building corals in shallow water contain zooxanthellae, and some deep water species, living at depths to which light cannot penetrate, form reefs but do not harbour the symbionts.

Staghorn coral (*Acropora cervicornis*) is an important hermatypic coral from the Caribbean

There are various types of shallow-water coral reef, including fringing reefs, barrier reefs and atolls; most occur in tropical and subtropical seas. They are very slow-growing, adding perhaps one centimetre (0.4 in) in height each year. The Great Barrier Reef is thought to have been laid down about two million years ago. Over time, corals fragment and die, sand and rubble accumulates between the corals, and the shells of clams and other molluscs decay to form a gradually evolving calcium carbonate structure. Coral reefs are extremely diverse marine ecosystems hosting over 4,000 species of fish, massive numbers of cnidarians, molluscs, crustaceans, and many other animals.

Evolutionary History

Although corals first appeared in the Cambrian period, some 542 million years ago, fossils are extremely rare until the Ordovician period, 100 million years later, when rugose and tabulate corals became widespread. Paleozoic corals often contained numerous endobiotic symbionts.

Solitary rugose coral (*Grewingkia*) in three views; Ordovician, southeastern Indiana

Tabulate corals occur in limestones and calcareous shales of the Ordovician and Silurian periods, and often form low cushions or branching masses of calcite alongside rugose corals. Their numbers began to decline during the middle of the Silurian period, and they became extinct at the end of the Permian period, 250 million years ago.

Rugose or horn corals became dominant by the middle of the Silurian period, and became extinct early in the Triassic period. The rugose corals existed in solitary and colonial forms, and were also composed of calcite.

The scleractinian corals filled the niche vacated by the extinct rugose and tabulate species. Their fossils may be found in small numbers in rocks from the Triassic period, and became common in the Jurassic and later periods. Scleractinian skeletons are composed of a form of calcium carbonate known as aragonite. Although they are geologically younger than the tabulate and rugose corals, the aragonite of their skeletons is less readily preserved, and their fossil record is accordingly less complete.

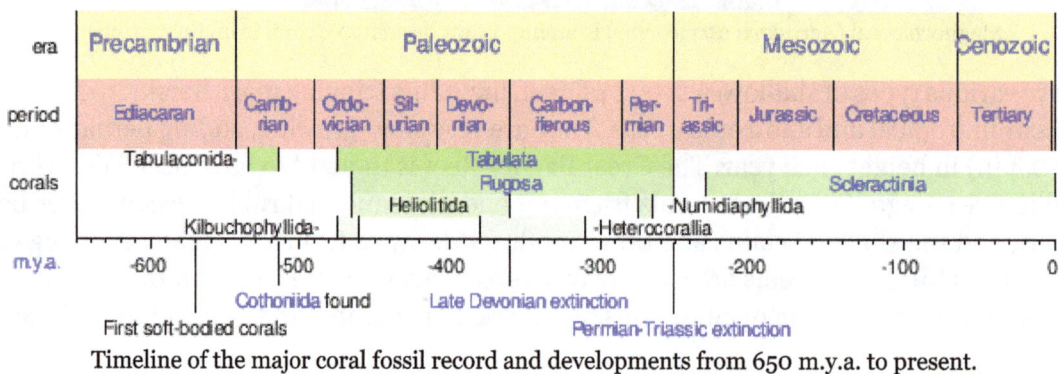

Timeline of the major coral fossil record and developments from 650 m.y.a. to present.

At certain times in the geological past, corals were very abundant. Like modern corals, these ancestors built reefs, some of which ended as great structures in sedimentary rocks. Fossils of fellow reef-dwellers algae, sponges, and the remains of many echinoids, brachiopods, bivalves, gastropods, and trilobites appear along with coral fossils. This makes some corals useful index fossils. Coral fossils are not restricted to reef remnants, and many solitary fossils may be found elsewhere, such as *Cyclocyathus*, which occurs in England's Gault clay formation.

Status

A healthy coral reef has a striking level of biodiversity in many forms of marine life.

Coral reefs are under stress around the world. In particular, coral mining, agricultural and urban runoff, pollution (organic and inorganic), overfishing, blast fishing, disease, and the digging of canals and access into islands and bays are localized threats to coral ecosystems. Broader threats are sea temperature rise, sea level rise and pH changes from ocean acidification, all associated with greenhouse gas emissions. In 1998, 16% of the world's reefs died as a result of increased water temperature.

Approximately 10% of the world's coral reefs are dead. About 60% of the world's reefs are at risk due to human-related activities. The threat to reef health is particularly strong in Southeast Asia, where 80% of reefs are endangered. Over 50% of the world's coral reefs may be destroyed by 2030; as a result, most nations protect them through environmental laws.

In the Caribbean and tropical Pacific, direct contact between ~40–70% of common seaweeds and coral causes bleaching and death to the coral via transfer of lipid-soluble metabolites. Seaweed and algae proliferate given adequate nutrients and limited grazing by herbivores such as parrotfish.

Water temperature changes of more than 1–2 °C (1.8–3.6 °F) or salinity changes can kill some species of coral. Under such environmental stresses, corals expel their Symbiodinium; without them coral tissues reveal the white of their skeletons, an event known as coral bleaching.

Submarine springs found along the coast of Mexico's Yucatán Peninsula produce water with a naturally low pH (relatively high acidity) providing conditions similar to those expected to become widespread as the oceans absorb carbon dioxide. Surveys discovered multiple species of live coral that appeared to tolerate the acidity. The colonies were small and patchily distributed, and had not formed structurally complex reefs such as those that compose the nearby Mesoamerican Barrier Reef System.

Protection

Marine Protected Areas (MPAs), Biosphere reserves, marine parks, national monuments world heritage status, fishery management and habitat protection can protect reefs from anthropogenic damage.

Many governments now prohibit removal of coral from reefs, and inform coastal residents about reef protection and ecology. While local action such as habitat restoration and herbivore protection can reduce local damage, the longer-term threats of acidification, temperature change and sea-level rise remain a challenge.

To eliminate destruction of corals in their indigenous regions, projects have been started to grow corals in non-tropical countries.

Relation to Humans

Local economies near major coral reefs benefit from an abundance of fish and other marine creatures as a food source. Reefs also provide recreational scuba diving and snorkeling tourism. These activities can damage coral but international projects such as Green Fins that encourage dive and snorkel centres to follow a Code of Conduct have been proven to mitigate these risks.

Live coral is highly sought after for aquaria. Soft corals are easier to maintain in captivity than hard corals.

Jewelry

Corals' many colors give it appeal for necklaces and other jewelry. Intensely red coral is prized as a gemstone. Sometimes called fire coral, it is not the same as fire coral. Red coral is very rare because of overharvesting.

6-strand necklace, Navajo (Native American), ca. 1920s, Brooklyn Museum

Always considered a precious mineral, "the Chinese have long associated red coral with auspiciousness and longevity because of its color and its resemblance to deer antlers (so by association, virtue, long life, and high rank". It reached its height of popularity during the Manchu or Qing Dynasty (1644-1911) when it was almost exclusively reserved for the emperor's use either in the form of coral beads (often combined with pearls) for court jewelry or as decorative Penjing (decorative miniature mineral trees). Coral was known as *shanhu* in Chinese. The "early-modern 'coral network' [began in] the Mediterranean Sea [and found its way] to Qing China via the English East India Company". There were strict rules regarding its use in a code established by the Qianlong Emperor in 1759.

Medicine

Depiction of coral in the Juliana Anicia Codex, a copy, written
in Constantinople in 515 AD, of Dioscorides' 1st century AD Greek work.

In medicine, chemical compounds from corals are used to treat cancer, AIDS and pain, and for other uses. Coral skeletons, e.g. *Isididae* are also used for bone grafting in humans. Coral Calx, known as Praval Bhasma in Sanskrit, is widely used in traditional system of Indian medicine as a supplement in the treatment of a variety of bone metabolic disorders associated with calcium deficiency. In classical times ingestion of pulverized coral, which consists mainly of the weak base calcium carbonate, was recommended for calming stomach ulcers by Galen and Dioscorides.

Construction

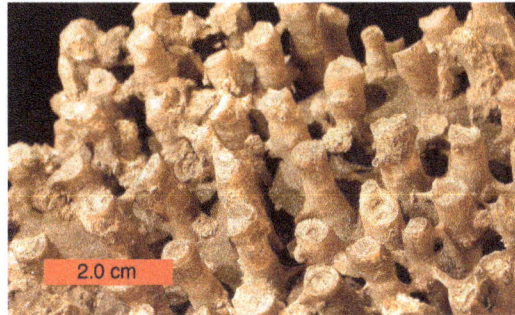

Tabulate coral (a syringoporid); Boone limestone (Lower Carboniferous) near Hiwasse, Arkansas, scale bar is 2.0 cm.

Coral reefs in places such as the East African coast are used as a source of building material. Ancient (fossil) coral limestone, notably including the Coral Rag Formation of the hills around Oxford (England), was once used as a building stone, and can be seen in some of the oldest buildings in that city including the Saxon tower of St Michael at the Northgate, St. George's Tower of Oxford Castle, and the mediaeval walls of the city.

Climate Research

Annual growth bands in some corals, such as the deep sea bamboo corals (*Isididae*), may be among the first signs of the effects of ocean acidification on marine life. The growth rings allow geologists to construct year-by-year chronologies, a form of incremental dating, which underlie high-resolution records of past climatic and environmental changes using geochemical techniques.

Certain species form communities called microatolls, which are colonies whose top is dead and mostly above the water line, but whose perimeter is mostly submerged and alive. Average tide level limits their height. By analyzing the various growth morphologies, microatolls offer a low resolution record of sea level change. Fossilized microatolls can also be dated using Radiocarbon dating. Such methods can help to reconstruct Holocene sea levels.

Increasing sea temperatures in tropical regions (~1 degree C) the last century have caused major coral bleaching, death, and therefore shrinking coral populations since although they are able to adapt and acclimate, it is uncertain if this evolutionary process will happen quickly enough to prevent major reduction of their numbers.

Though coral have large sexually-reproducing populations, their evolution can be slowed by abundant asexual reproduction. Gene flow is variable among coral species. According to the biogeography of coral species gene flow cannot be counted on as a dependable source of adaptation as they are very stationary organisms. Also, coral longevity might factor into their adaptivity.

However, adaptation to climate changes has been demonstrated in many cases. These are usually due to a shift in coral and zooxanthellae genotypes. These shifts in allelic frequencies have progressed toward more tolerant types of zooxanthellae. Scientists found that a certain scleractinian zooxanthella is becoming more common where sea temperature is high. Symbionts able to tolerate warmer water seem to photosynthesise more slowly, implying an evolutionary trade-off.

In the Gulf of Mexico, where sea temperatures are rising, cold-sensitive staghorn and elkhorn coral have shifted in location. Not only have the symbionts and specific species been shown to shift, but there seems to be a certain growth rate favorable to selection. Slower-growing but more heat-tolerant corals have become more common. The changes in temperature and acclimation are complex. Some reefs in current shadows represent a refugium location that will help them adjust to the disparity in the environment even if eventually the temperatures may rise more quickly there than in other locations. This separation of populations by climatic barriers causes a realized niche to shrink greatly in comparison to the old fundamental niche.

Geochemistry

Corals are shallow, colonial organisms that integrate $\delta^{18}O$ and trace elements into their skeletal aragonite (polymorph of calcite) crystalline structures, as they grow. Geochemistry anomalies within the crystalline structures of corals represent functions of temperature, salinity and oxygen isotopic composition. Such geochemical analysis can help with climate modeling.

Strontium/Calcium Ratio Anomaly

Time can be attributed to coral geochemistry anomalies by correlating strontium/calcium minimums with sea surface temperature (SST) maximums to data collected from NINO 3.4 SSTA.

Oxygen Isotope Anomaly

The comparison of coral strontium/calcium minimums with sea surface temperature maximums, data recorded from NINO 3.4 SSTA, time can be correlated to coral strontium/calcium and $\delta^{18}O$ variations. To confirm accuracy of the annual relationship between Sr/Ca and $\delta^{18}O$ variations, a perceptible association to annual coral growth rings confirms the age conversion. Geochronology is established by the blending of Sr/Ca data, growth rings, and stable isotope data. El Nino-Southern Oscillation (ENSO) is directly related to climate fluctuations that influence coral $\delta^{18}O$ ratio from local salinity variations associated with the position of the South Pacific convergence zone (SPCZ) and can be used for ENSO modeling.

Sea Surface Temperature and Sea Surface Salinity

The global moisture budget is primarily being influenced by tropical sea surface temperatures from the position of the Intertropical Convergence Zone (ITCZ). The Southern Hemisphere has a unique meteorological feature positioned in the southwestern Pacific Basin called the South Pacific Convergence Zone (SPCZ), which contains a perennial position within the Southern Hemisphere. During ENSO warm periods, the SPCZ reverses orientation extending from the equator down south through Solomon Islands, Vanuatu, Fiji and towards the French Polynesian Islands;

and due east towards South America affecting geochemistry of corals in tropical regions.

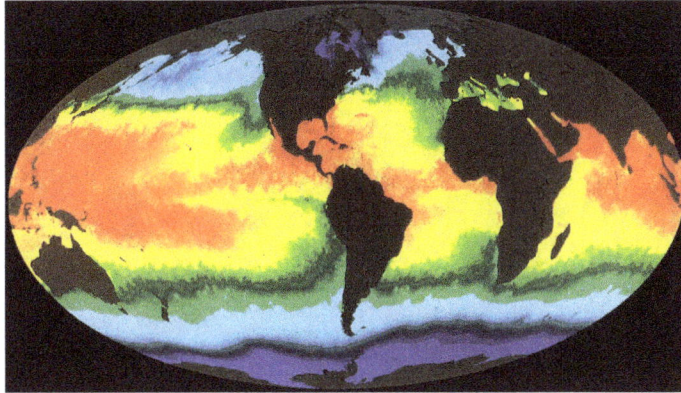

Global sea surface temperature (SST)

Geochemical analysis of skeletal coral can be linked to sea surface salinity (SSS) and sea surface temperature (SST), from El Nino 3.4 SSTA data, of tropical oceans to seawater $\delta^{18}O$ ratio anomalies from corals. ENSO phenomenon can be related to variations in sea surface salinity (SSS) and sea surface temperature (SST) that can help model tropical climate activities.

Limited Climate Research on Current Species

Climate research on live coral species is limited to a few studied species. Studying *Porites* coral provides a stable foundation for geochemical interpretations that is much simpler to physically extract data in comparison to *Platygyra* species where the complexity of *Platygyra* species skeletal structure creates difficulty when physically sampled, which happens to be one of the only multidecadal living coral records used for coral paleoclimate modeling.

Genus: *Porites lutea*

Aquaria

The saltwater fishkeeping hobby has increasingly expanded, over recent years, to include reef tanks, fish tanks that include large amounts of live rock on which coral is allowed to grow and spread. These tanks are either kept in a natural-like state, with algae (sometimes in the form of an algae scrubber) and a deep sand bed providing filtration, or as "show tanks", with the rock kept largely bare of the algae and microfauna that would normally populate it, in order to appear neat and clean.

This dragon-eye zoanthid is a popular source of color in reef tanks

The most popular kind of coral kept is soft coral, especially zoanthids and mushroom corals, which are especially easy to grow and propagate in a wide variety of conditions, because they originate in enclosed parts of reefs where water conditions vary and lighting may be less reliable and direct. More serious fishkeepers may keep small polyp stony coral, which is from open, brightly lit reef conditions and therefore much more demanding, while large polyp stony coral is a sort of compromise between the two.

Aquaculture

Coral aquaculture, also known as *coral farming* or *coral gardening*, is the cultivation of corals for commercial purposes or coral reef restoration. Aquaculture is showing promise as a potentially effective tool for restoring coral reefs, which have been declining around the world. The process bypasses the early growth stages of corals when they are most at risk of dying. Coral fragments known as "seeds" are grown in nurseries then replanted on the reef. Coral is farmed by coral farmers who live locally to the reefs and farm for reef conservation or for income. It is also farmed by scientists for research, by businesses for the supply of the live and ornamental coral trade and by private aquarium hobbyists.

References

- Kuraku; et al. (December 1999). "Monophyly of Lampreys and Hagfishes Supported by Nuclear DNA–Coded Genes". Journal of Molecular Evolution. 49 (6): 729–35. PMID 10594174. doi:10.1007/PL00006595

- Liem, K.F.; Walker, W.F. (2001). Functional anatomy of the vertebrates: an evolutionary perspective. Harcourt College Publishers. p. 277. ISBN 978-0-03-022369-3

- Frost, Darrel (2013). "American Museum of Natural History: Amphibian Species of the World 5.6, an Online Reference". The American Museum of Natural History. Retrieved October 24, 2013

- Shu; et al. (4 November 1999). "Lower Cambrian vertebrates from south China". Nature. 402 (6757): 42–46. Bibcode:1999Natur.402...42S. doi:10.1038/46965

- Lombard, R. E.; Bolt, J. R. (1979). "Evolution of the tetrapod ear: an analysis and reinterpretation". Biological Journal of the Linnean Society. 11 (1): 19–76. doi:10.1111/j.1095-8312.1979.tb00027.x

- Andersen, N.M.; Weir, T.A. (2004). Australian water bugs: their biology and identification (Hemiptera-Heteroptera, Gerromorpha & Nepomorpha). Apollo Books. p. 38. ISBN 978-87-88757-78-1

- Kouchinsky, A. (2000). "Shell microstructures in Early Cambrian molluscs" (PDF). Acta Palaeontologica Polonica. 45 (2): 119–150. Retrieved 4 Nov 2009

- Beneski, John T. Jr. (September 1989). "Adaptive significance of tail autotomy in the salamander, Ensatina". Journal of Herpetology. 23 (3): 322–324. JSTOR 1564465. doi:10.2307/1564465

- Moyle, Peter B.; Cech, Joseph J. (2003). Fishes, An Introduction to Ichthyology (5th ed.). Benjamin Cummings. ISBN 978-0-13-100847-2

- Mallet, Jim (12 June 2007). "Taxonomy of Lepidoptera: the scale of the problem". The Lepidoptera Taxome Project. University College, London. Retrieved 8 February 2011

- Baird, Donald (May 1965). "Paleozoic lepospondyl amphibians". Integrative and Comparative Biology. 5 (2): 287–294. doi:10.1093/icb/5.2.287

- Waldman, Bruce; Ryan, Michael J. (1983). "Thermal advantages of communal egg mass deposition in wood frogs (Rana sylvatica)". Journal of Herpetology. 17 (1): 70–72. JSTOR 1563783. doi:10.2307/1563783

- Roček, Z. (2000). "14. Mesozoic Amphibians". In Heatwole, H.; Carroll, R. L. Amphibian Biology: Paleontology: The Evolutionary History of Amphibians (PDF). 4. Surrey Beatty & Sons. pp. 1295–1331. ISBN 978-0-949324-87-0

- Hero, Jean-Marc; Clarke, John; Meyer, Ed (2004). "Assa darlingtoni". IUCN Red List of Threatened Species. Version 2012.2. Retrieved November 20, 2012.

- Brainerd, E. L. (1999). "New perspectives on the evolution of lung ventilation mechanisms in vertebrates". Experimental Biology Online. 4 (2): 1–28. doi:10.1007/s00898-999-0002-1

- Breckenridge, W. R.; Nathanael, S.; Pereira, L. (1987). "Some aspects of the biology and development of Ichthyophis glutinosus". Journal of Zoology. 211: 437–449

- Sumich, James L.; Morrissey, John F. (2004). Introduction to the Biology of Marine Life. Jones & Bartlett Learning. p. 171. ISBN 978-0-7637-3313-1

- Duellman, William E.; Zug, George R. (2012). "Amphibian". Encyclopædia Britannica Online. Encyclopædia Britannica. Retrieved March 27, 2012

- Crump, Martha L. (1986). "Cannibalism by younger tadpoles: another hazard of metamorphosis". Copeia. 4 (4): 1007–1009. JSTOR 1445301. doi:10.2307/1445301.

- Trueb, Linda; Gans, Carl (1983). "Feeding specializations of the Mexican burrowing toad, Rhinophrynus dorsalis (Anura: Rhinophrynidae)". Journal of Zoology. 199 (2): 189–208. doi:10.1111/j.1469-7998.1983.tb02090.x

- Crump, Martha L. (1996). "Parental care among the Amphibia". Advances in the Study of Behavior. Advances in the Study of Behavior. 25: 109–144. ISBN 978-0-12-004525-9. doi:10.1016/S0065-3454(08)60331-9

Animal Classification on the basis of Eating Patterns

Animal nutrition and diet studies the nutrition that animals require to stay healthy. Plants possess autotro-phic mode of nutrition while animals need to consume food for nutrition. This chapter will provide an integrated understanding of animal nutrition.

Animal Nutrition

A juvenile Red-tailed Hawk eating a California Vole

Animal nutrition focuses on the dietary needs of domesticated animals, primarily those in agriculture and food production.

There are seven major classes of nutrients; carbohydrates, fats, fibre, minerals, protein, vitamin, and water.

The macronutrients (excluding fiber and water) provide structural material (amino acids from which proteins are built, and lipids from which cell membranes and some signaling molecules are built) and energy. Some of the structural material can be used to generate energy internally, and in either case it is measured in joules or calories (sometimes called "kilocalories" and on other rare occasions written with a capital C to distinguish them from little 'c' calories). Carbohydrates and proteins provide 17 kJ approximately (4 kcal) of energy per gram, while fats provide 37 kJ (9 kcal) per gram., though the net energy from either depends on such factors as absorption and digestive effort, which vary substantially from instance to instance. Vitamins, minerals, fiber, and water do not provide energy, but are required for other reasons. A third class dietary material, fiber (i.e., non-digestible material such as cellulose), seems also to be required, for both mechanical and bio-chemical reasons, though the exact reasons remain unclear.

Molecules of carbohydrates and fats consist of carbon, hydrogen, and oxygen atoms. Carbohydrates range from simple monosaccharides (glucose, fructose, galactose) to complex polysaccharides (starch). Fats are triglycerides, made of assorted fatty acid monomers bound to glycerol backbone. Some fatty acids, but not all, are essential in the diet they cannot be synthesized in the body. Protein molecules contain nitrogen atoms in addition to carbon, oxygen, and hydrogen. The fundamental components of protein are nitrogen-containing amino acids, some of which are essential in the sense that humans cannot make them internally. Some of the amino acids are convertible (with the expenditure of energy) to glucose and can be used for energy production just as ordinary glucose. By breaking down existing protein, some glucose can be produced internally; the remaining amino acids are discarded, primarily as urea in urine. This occurs normally only during prolonged starvation.

Other micronutrients include antioxidants and phytochemicals which are said to influence (or protect) some body systems. Their necessity is not as well established as in the case of, for instance, vitamins.

Most foods contain a mix of some or all of the nutrient classes, together with other substances such as toxins or various sorts. Some nutrients can be stored internally (e.g., the fat soluble vitamins), while others are required more or less continuously. Poor health can be caused by a lack of required nutrients or, in extreme cases, too much of a required nutrient. For example, both salt and water (both absolutely required) will cause illness or even death in too large amounts.

Heterotrophs and Autotrophs

Autotrophs~ Those organisms which can make food on their own, from simple substances , are known as autotrophs. This is because green plants chlorophyll and can perform photosynthesis. In other words, green plants have autotrophic mode of nutrition.

Heterotrophs~ The organisms which cannot make food themselves by the process of photosynthesis and take food from green plants or animals, are called heterotrophs. All non-green plants do not have chlorophyll for carrying the process of photosynthesis. So, they depend other organisms for obtaining food.

Fat

A molecule of dietary fat typically consists of several fatty acids (containing long chains of carbon and hydrogen atoms), bonded to a glycerol. They are typically found as triglycerides (three fatty acids attached to one glycerol backbone). Fats may be classified as saturated or unsaturated depending on the detailed structure of the fatty acids involved. Saturated fats have all of the carbon atoms! in their fatty acid chains bonded to hydrogen atoms, whereas unsaturated fats have some of these carbon atoms double-bonded, so their molecules have relatively fewer hydrogen atoms than a saturated fatty acid of the same length. Unsaturated fats may be further classified as monounsaturated (one double-bond) or polyunsaturated (many double-bonds). Furthermore, depending on the location of the double-bond in the fatty acid chain, unsaturated fatty acids are classified as omega-3 or omega-6 fatty acids. Trans fats are a type of unsaturated fat with *trans*-isomer bonds; these are rare in nature and in foods from natural sources; they are typically created in an industrial process called (partial) hydrogenation.

Many studies have shown that unsaturated fats, particularly monounsaturated fats, are best in the human diet. Saturated fats, typically from animal sources, are next, while trans fats are to be avoided.

Saturated and some trans fats are typically solid at room temperature (such as butter or lard), while unsaturated fats are typically liquids (such as olive oil or flaxseed oil). Trans fats are very rare in nature, but have properties useful in the food processing industry, such as rancid resistance.

Essential Fatty Acids

Most fatty acids are non-essential, meaning the body can produce them as needed, generally from other fatty acids and always by expending energy to do so. However, in humans at least two fatty acids are essential and must be included in the diet. An appropriate balance of essential fatty acids such as omega-3 and omega-6, fatty acids seems also important for health, though definitive experimental demonstration has been elusive. Both of these "omega" long-chain polyunsaturated fatty acids are substrates for a class of eicosanoids known as prostaglandins, which have roles throughout the human body. They are hormones, in some respects. The omega-3 eicosapentaenoic acid (EPA), which can be made in the human body from the omega-3 essential fatty acid alpha-linolenic acid (LNA), or taken in through marine food sources, serves as a building block for series 3 prostaglandins (e.g. weakly inflammatory PGE3). The omega-6 dihomo-gamma-linolenic acid (DGLA) serves as a building block for series 1 prostaglandins (e.g. anti-inflammatory PGE1), whereas arachidonic acid (AA) serves as a building block for series 2 prostaglandins (e.g. pro-inflammatory PGE 2). Both DGLA and AA can be made from the omega-6 linoleic acid (LA) in the human body, or can be taken in directly through food. An appropriately balanced intake of omega-3 and omega-6 partly determines the relative production of different prostaglandins; one reason a balance between omega-3 and omega-6 is believed important for cardiovascular health. In industrialized societies, people typically consume large amounts of processed vegetable oils, which have reduced amounts of the essential fatty acids along with too much of omega-6 fatty acids relative to omega-3 fatty acids.

The conversion rate of omega-6 DGLA to AA largely determines the production of the prostaglandins PGE1 and PGE2. Omega-3 EPA prevents AA from being released from membranes, thereby skewing prostaglandin balance away from pro-inflammatory PGE2 (made from AA) toward anti-inflammatory PGE1 (made from DGLA). Moreover, the conversion (desaturation) of DGLA to AA is controlled by the enzyme delta-5-desaturase, which in turn is controlled by hormones such as insulin (up-regulation) and glucagon (down-regulation). The amount and type of carbohydrates consumed, along with some types of amino acid, can influence processes involving insulin, glucagon, and other hormones; therefore the ratio of omega-3 versus omega-6 has wide effects on general health, and specific effects on immune function and inflammation, and mitosis (i.e. cell division).

Good sources of essential fatty acids include most vegetables, nuts, seeds, and marine oils, Some of the best sources are fish, flax seed oils, soy beans, pumpkin seeds, sunflower seeds, and walnuts.

Fiber

Dietary fiber is a carbohydrate (polysaccharide or oligosaccharide) that is incompletely absorbed in humans and in some animals. Like all carbohydrates, when it is metabolized it can produce four calories (kilocalories) of energy per gram. But in most circumstances, it accounts for less than that because of its limited absorption and digestibility. Dietary fiber consists mainly of cellulose, a large carbohydrate polymer that is indigestible because humans do not have the required enzymes to disassemble it. There are two subcategories; soluble and insoluble fibre. Whole grains, fruits (especially plums, prunes, and figs), and vegetables are good sources of dietary fiber. Fiber is important to digestive health and is thought to reduce the risk of colon cancer. For mechanical reasons it can help in alleviating both consti-pation and diarrhea. Fiber provides bulk to the intestinal contents, and insoluble fiber especially stim-ulates peristalsis—the rhythmic muscular contractions of the intestines which move digesta along the digestive tract. Some soluble fibers produce a solution of high viscosity; this is essentially a gel, which slows the movement of food through the intestines. Additionally, fiber, perhaps especially that from whole grains, may help lessen insulin spikes and reduce the risk of type 2 diabetes.

Protein

Most meats such as chicken contain all the essential amino acids needed for humans

Proteins are the basis of many animal body structures (e.g. muscles, skin, and hair). They also form the enyzmes which control chemical reactions throughout the body. Each molecule is composed of amino acids which are characterized by inclusion of nitrogen and sometimes sulphur (these com-ponents are responsible for the distinctive smell of burning protein, such as the keratin in hair). The body requires amino acids to produce new proteins (protein retention) and to replace damaged proteins (maintenance). As there is no protein or amino acid storage provision, amino acids must be present in the diet. Excess amino acids are discarded, typically in the urine. For all animals, some amino acids are *essential* (an animal cannot produce them internally) and some are *non-essential* (the animal can produce them from other nitrogen-containing compounds). About twenty amino ac-ids are found in the human body, and about ten of these are essential, and therefore must be included in the diet. A diet that contains adequate amounts of amino acids (especially those that are essential) is particularly important in some situations; during early development and maturation, pregnancy, lactation, or injury (a burn, for instance). A *complete* protein source contains all the essential amino acids; an *incomplete* protein source lacks one or more of the essential amino acids.

It is possible to combine two incomplete protein sources (e.g. rice and beans) to make a complete protein source, and characteristic combinations are the basis of distinct cultural cooking tradi-tions. Sources of dietary protein include meats, tofu and other soy-products, eggs, grains, legumes, and dairy products such as milk and cheese. A few amino acids from protein can be converted into glucose and used for fuel through a process called gluconeogenesis; this is done in quantity only during starvation. The amino acids remaining after such conversion are discarded.

Minerals

Dietary minerals are the chemical elements required by living organisms, other than the four ele-ments carbon, hydrogen, nitrogen, and oxygen that are present in nearly all organic molecules. The term "mineral" is archaic, since the intent is to describe simply the less common elements in the diet.

Some are heavier than the four just mentioned—including several metals, which often occur as ions in the body. Some dietitians recommend that these be supplied from foods in which they occur naturally, or at least as complex compounds, or sometimes even from natural inorganic sources (such as calcium carbonate from ground oyster shells). Some are absorbed much more readily in the ionic forms found in such sources. On the other hand, minerals are often artificially added to the diet as supplements; the most famous is likely iodine in iodized salt which prevents goiter.

Macrominerals

Many elements are essential in relative quantity; they are usually called "bulk minerals". Some are structural, but many play a role as electrolytes. Elements with recommended dietary allowance (RDA) greater than 200 mg/day are, in alphabetical order (with informal or folk-medicine perspectives in parentheses).

- Calcium, a common electrolyte, but also needed structurally structural (for muscle and digestive system health, bones, some forms neutralizes acidity, may help clear toxins, and provide signaling ions for nerve and membrane functions)

- Chlorine as chloride ions; very common electrolyte

- Magnesium, required for processing ATP and related reactions (builds bone, causes strong peristalsis, increases flexibility, increases alkalinity)

- Phosphorus, required component of bones; essential for energy processing

- Potassium, a very common electrolyte (heart and nerve health)

- Sodium, a very common electrolyte; not generally found in dietary supplements, despite being needed in large quantities, because the ion is very common in food; typically as sodium chloride, or common salt

- Sulfur for three essential amino acids and therefore many proteins (skin, hair, nails, liver, and pancreas)

Trace Minerals

Many elements are required in trace amounts, usually because they play a catalytic role in enzymes. Some trace mineral elements (RDA < 200 mg/day) are, in alphabetical order:

- Cobalt required for biosynthesis of vitamin B12 family of coenzymes

- Copper required component of many redox enzymes, including cytochrome c oxidase

- Chromium required for sugar metabolism

- Iodine required not only for the biosynthesis of thyroxin, but probably, for other important organs as breast, stomach, salivary glands, thymus etc.; for this reason iodine is needed in larger quantities than others in this list, and sometimes classified with the macrominerals

- Iron required for many enzymes, and for hemoglobin and some other proteins

- Manganese (processing of oxygen)

- Molybdenum required for xanthine oxidase and related oxidases

- Nickel present in urease

- Selenium required for peroxidase (antioxidant proteins)

- Vanadium (Speculative: there is no established RDA for vanadium. No specific biochemical function has been identified for it in humans, although vanadium is required for some lower organisms.)

- Zinc required for several enzymes such as carboxypeptidase, liver alcohol dehydrogenase, carbonic anhydrase

Vitamins

As with the minerals discussed above, some vitamins are recognized as essential nutrients, necessary in the diet for good health. (Vitamin D is the exception; it can alternatively be synthesized in the skin, in the presence of UVB radiation.) Certain vitamin-like compounds that are recommended in the diet, such as carnitine, are thought useful for survival and health, but these are not "essential" dietary nutrients because the human body has some capacity to produce them from other compounds. Moreover, thousands of different phytochemicals have recently been discovered in food (particularly in fresh vegetables), which may have desirable properties including antioxidant activity; experimental demonstration has been suggestive but inconclusive. Other essential nutrients not classed as vitamins include essential amino acids, choline, essential fatty acids, and the minerals.

Vitamin deficiencies may result in disease conditions: goitre, scurvy, osteoporosis, impaired immune system, disorders of cell metabolism, certain forms of cancer, symptoms of premature aging, and poor psychological health (including eating disorders), among many others. Excess of some vitamins is also dangerous to health (notably vitamin A), and animal nutrition researchers have managed to establish safe levels for some common companion animals. For at least one vitamin, B6, toxicity begins at levels not far above the required amount. Deficiency or excess of minerals can also have serious health consequences.

Water

A manual water pump in China

About 70% of the non-fat mass of the human body is made of water. Analysis of Adipose Tissue in Relation to Body Weight Loss in Man. Retrieved from Journal of Applied To function properly, the body requires between one and seven liters of water per day to avoid dehydration; the precise amount depends on the level of activity, temperature, humidity, and other factors. With physical exertion and heat exposure, water loss increases and daily fluid needs will eventually increase as well.

It is not fully clear how much water intake is needed by healthy people, although some experts assert that 8–10 glasses of water (approximately 2 liters) daily is the minimum to maintain proper hydration. The notion that a person should consume eight glasses of water per day cannot be traced to a credible scientific source. The effect of, greater or lesser, water intake on weight loss and on constipation is also still unclear. The original water intake recommendation in 1945 by the Food and Nutrition Board of the National Research Council read: "An ordinary standard for diverse persons is 1 milliliter for each calorie of food. Most of this quantity is contained in prepared foods." The latest dietary reference intake report by the United States National Research Council recommended, generally, (including food sources); 2.7 liters of water total for women and 3.7 liters for men. Specifically, pregnant and breastfeeding women need additional fluids to stay hydrated. According to the Institute of Medicine—who recommend that, on average, women consume 2.2 litres and men 3.0 litres—this is recommended to be 2.4 litres (approx. 9 cups) for pregnant women and 3 litres (approx. 12.5 cups) for breastfeeding women since an especially large amount of fluid is lost during nursing.

People can drink far more water than necessary while exercising, however, putting them at risk of water intoxication, which can be fatal. In particular large amounts of de-ionized water are dangerous.

Normally, about 20 percent of water intake comes in food, while the rest comes from drinking water and assorted beverages (caffeinated included). Water is excreted from the body in multiple forms; including urine and feces, sweating, and by water vapor in the exhaled breath.

Other Nutrients

Other micronutrients include antioxidants and phytochemicals. These substances are generally more recent discoveries which have not yet been recognized as vitamins or as required. Phytochemicals may act as antioxidants, but not all phytochemicals are antioxidants.

Antioxidants

Antioxidants are a recent discovery. As cellular metabolism/energy production requires oxygen, potentially damaging (e.g. mutation causing) compounds known as free radicals can form. Most of these are oxidizers (i.e. acceptors of electrons) and some react very strongly. For normal cellular maintenance, growth, and division, these free radicals must be sufficiently neutralized by antioxidant compounds. Some are produced by the human body with adequate precursors (glutathione, Vitamin C) and those the body cannot produce may only be obtained in the diet via direct sources (Vitamin C in humans, Vitamin A, Vitamin K) or produced by the body from other compounds (Beta-carotene converted to Vitamin A by the body, Vitamin D synthesized from cholesterol by sunlight). Phytochemicals and their subgroup polyphenols are the majority of antioxidants; about 4,000 are known. Different antioxidants are now known to function in a cooperative network, e.g. vitamin C can reactivate free radical-containing glutathione or vitamin E by accepting the free radical itself, and so on. Some antioxidants are more effective than others at neutralizing different free radicals. Some cannot neutralize certain free radicals. Some cannot be present in certain areas of free radical development (Vitamin A is fat-soluble and protects fat areas, Vitamin C is water soluble and protects those areas). When interacting with a free radical, some antioxidants produce a different free radical compound that is less dangerous or more dangerous than the previous compound. Having a variety of antioxidants allows any byproducts to be safely dealt with by more efficient antioxidants in neutralizing a free radical's butterfly effect.

Phytochemicals

Blackberries are a source of polyphenol antioxidants

A growing area of interest is the effect upon human health of trace chemicals, collectively called phytochemicals. These nutrients are typically found in edible plants, especially colorful fruits and vegetables, but also other organisms including seafood, algae, and fungi. The effects of phytochemicals increasingly survive rigorous testing by prominent health organizations. One of the principal classes of phytochemicals are polyphenol antioxidants, chemicals which are known to provide certain health benefits to the cardiovascular system and immune system. These chemicals are known to down-regulate the formation of reactive oxygen species, key chemicals in cardiovascular disease.

Perhaps the most rigorously tested phytochemical is zeaxanthin, a yellow-pigmented carotenoid present in many yellow and orange fruits and vegetables. Repeated studies have shown a strong correlation between ingestion of zeaxanthin and the prevention and treatment of age-related macular degeneration (AMD). Less rigorous studies have proposed a correlation between zeaxanthin intake and cataracts. A second carotenoid, lutein, has also been shown to lower the risk of contracting AMD. Both compounds have been observed to collect in the retina when ingested orally, and they serve to protect the rods and cones against the destructive effects of light.

Another carotenoid, beta-cryptoxanthin, appears to protect against chronic joint inflammatory diseases, such as arthritis. While the association between serum blood levels of beta-cryptoxanthin and substantially decreased joint disease has been established, neither a convincing mechanism for such protection nor a cause-and-effect have been rigorously studied. Similarly, a red phytochemical, lycopene, has substantial credible evidence of negative association with development of prostate cancer.

The correlations between the ingestion of some phytochemicals and the prevention of disease are, in some cases, enormous in magnitude.

Even when the evidence is obtained, translating it to practical dietary advice can be difficult and counter-intuitive. Lutein, for example, occurs in many yellow and orange fruits and vegetables and protects the eyes against various diseases. However, it does not protect the eye nearly as well as zeaxanthin, and the presence of lutein in the retina will prevent zeaxanthin uptake. Additionally, evidence has shown that the lutein present in egg yolk is more readily absorbed than the lutein from vegetable sources, possibly because of fat solubility. At the most basic level, the question "should you eat eggs?" is complex to the point of dismay, including misperceptions about the health effects of cholesterol in egg yolk, and its saturated fat content.

As another example, lycopene is prevalent in tomatoes (and actually is the chemical that gives tomatoes their red color). It is more highly concentrated, however, in processed tomato products such as commercial pasta sauce, or tomato soup, than in fresh "healthy" tomatoes. Yet, such sauces tend to have high amounts of salt, sugar, other substances a person may wish or even need to avoid.

The following table presents phytochemical groups and common sources, arranged by family:

Family	Sources	Possible Benefits
flavonoids	berries, herbs, vegetables, wine, grapes, tea	general antioxidant, oxidation of LDLs, prevention of arteriosclerosis and heart disease
isoflavones (phytoestrogens)	soy, red clover, kudzu root	general antioxidant, prevention of arteriosclerosis and heart disease, easing symptoms of menopause, cancer prevention
isothiocyanates	cruciferous vegetables	cancer prevention
monoterpenes	citrus peels, essential oils, herbs, spices, green plants, atmosphere	cancer prevention, treating gallstones
organosulfur compounds	chives, garlic, onions	cancer prevention, lowered LDLs, assistance to the immune system
saponins	beans, cereals, herbs	Hypercholesterolemia, Hyperglycemia, Antioxidant, cancer prevention, Anti-inflammatory
capsaicinoids	all *capiscum* (chile) peppers	topical pain relief, cancer prevention, cancer cell apoptosis

Ash

Though not really a nutrient as such, an entry for *ash* is sometimes found on nutrition labels, especially for pet food. This entry measures the weight of inorganic material left over after the food is burned for two hours at 600 °C. Thus, it does not include water, fibre, and nutrients that provide calories, but it does include some nutrients, such as minerals.

There have been some concerns that too much ash may contribute to feline urological syndrome in domestic cats.

Intestinal Bacterial Flora

It is now also known that animal intestines contain a large population of gut flora. In humans, these include species such as *Bacteroides*, *L. acidophilus* and *E. coli*, among many others. They are essential to digestion, and are also affected by the food we eat. Bacteria in the gut perform many important functions for humans, including breaking down and aiding in the absorption of otherwise indigestible food; stimulating cell growth; repressing the growth of harmful bacteria, training the immune system to respond only to pathogens; producing vitamin B12, and defending against some infectious diseases.

Carnivore

Lion eating meat

A carnivore meaning 'Meat Eater' (Latin, *caro* meaning 'meat' or 'flesh' and *vorare* meaning 'to devour') is an organism that derives its energy and nutrient requirements from a diet consisting mainly or exclusively of animal tissue, whether through predation or scavenging. Animals that depend solely on animal flesh for their nutrient requirements are called obligate carnivores while those that also consume non-animal food are called facultative carnivores. Omnivores also consume both animal and non-animal food, and apart from the more general definition, there is no clearly defined ratio of plant to animal material that would distinguish a facultative carnivore from an omnivore. A carnivore that sits at the top of the food chain is termed an apex predator.

Plants that capture and digest insects (and, at times, other small animals) are called carnivorous plants. Similarly, fungi that capture microscopic animals are often called carnivorous fungi.

Classification

The word "carnivore" sometimes refers to the mammalian order Carnivora, but this is somewhat misleading. While many Carnivora meet the definition of being meat eaters, not all do, and even fewer are true obligate carnivores. For example, most species of bears are actually omnivorous, except for the giant panda, which is almost exclusively herbivorous, and the exclusively meat-eating polar bear, which lives in the Arctic, where few plants grow. In addition, there are plenty of carnivorous species that are not members of Carnivora.

Outside the animal kingdom, there are several genera containing carnivorous plants and several phyla containing carnivorous fungi. The former are predominantly insectivores, while the latter prey mostly on microscopic invertebrates, such as nematodes, amoebae and springtails.

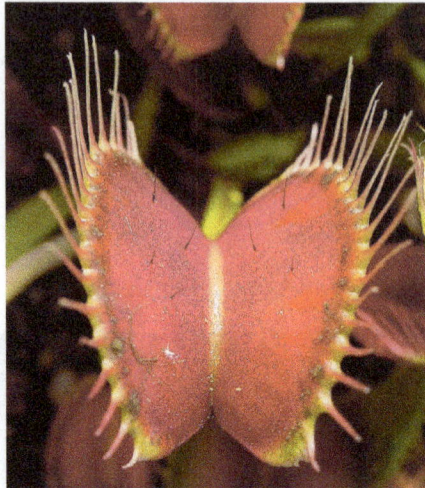

The Venus flytrap, a well known carnivorous plant

Carnivores are sometimes characterized by the type of prey that they consume. For example, animals that eat insects and similar invertebrates primarily or exclusively are called insectivores, while those that eat fish primarily or exclusively are called piscivores. The first tetrapods, or land-dwelling vertebrates, were piscivorous amphibians known as labyrinthodonts. They gave rise to insectivorous vertebrates and, later, to predators of other tetrapods.

Carnivores may alternatively be classified according to the percentage of meat in their diet. The diet of a hypercarnivore consists of more than 70% meat, that of a mesocarnivore 50–70%, and that of a hypocarnivore less than 30%, with the balance consisting of non-animal foods such as fruits, other plant material, or fungi.

Obligate Carnivores

This Bengal tiger's sharp teeth and strong jaws are the classical physical traits expected
from carnivorous mammalian predators

Obligate carnivores, or "true" carnivores, are those carnivores whose survival depends on nutrients which are found only in animal flesh. While obligate carnivores might be able to ingest small amounts of plant material, because of their evolution they lack the necessary physiology required to digest that plant matter. In fact, some obligate carnivorous mammals will only ever ingest vegetation for its specific use as an emetic to self-induce vomiting to rid itself of food that has upset its stomach.

For instance, felids including the domestic cat are obligate carnivores requiring a diet of primarily animal flesh and organs. Specifically, cats have high protein requirements and their metabolisms appear unable to synthesize certain essential nutrients (including retinol, arginine, taurine, and arachidonic acid), and thus, in nature, they can rely only on animal flesh as their diet to supply these nutrients.

Synthetic man-made forms of essential nutrients which in nature are found only in flesh, such as taurine, have been developed in labs thanks to modern human scientific achievements, and this has allowed feed manufacturers to formulate foods for carnivores, including domestic pets and zoo animals, with varying amounts of plant material, largely for cost effectiveness. The diets of these domesticated carnivores and carnivores in human captivity, however, must nevertheless always consist of mostly naturally-derived flesh feed. Imposing diets which are largely, mostly, or entirely herbivorous (or even omnivorous) on obligate carnivores (like raising "vegan" housebound domestic cats which are prevented from hunting their own prey outside) will lead to malnutrition and other serious (many times fatal) health conditions.

Characteristics of Carnivores

Characteristics commonly associated with carnivores include organs for capturing and disarticulating prey (teeth and claws serve these functions in many vertebrates) and status as a predator. In truth, these assumptions may be misleading, as some carnivores do not hunt and are scavengers (though most hunting carnivores will scavenge when the opportunity exists). Thus they do not have the characteristics associated with hunting carnivores. Carnivores have comparatively short digestive systems, as they are not required to break down tough cellulose found in plants. Many animals that hunt other animals have evolved eyes that face forward, thus making depth perception possible. This is almost universal among mammalian predators. Other predators, like crocodiles, as well as most reptiles and amphibians, have sideways facing eyes and hunt by ambush rather than pursuit.

Prehistoric Carnivores

The first vertebrate carnivores were fish, and then amphibians that moved on to land. Early tetrapods were large amphibious piscivores. Some scientists assert that *Dimetrodon* "was the first terrestrial vertebrate to develop the curved, serrated teeth that enable a predator to eat prey much larger than itself." While amphibians continued to feed on fish and later insects, reptiles began exploring two new food types, tetrapods (carnivory), and later, plants (herbivory). Carnivory was a natural transition from insectivory for medium and large tetrapods, requiring minimal adaptation (in contrast, a complex set of adaptations was necessary for feeding on highly fibrous plant materials).

Carnivoramorphs are currently the dominant carnivorous mammals, and have been so since the Miocene. In the early to mid-Cenozoic, however, hyaenodonts, oxyaenid, entelodonts, ptolemaiidans, "arctocyonids" and "mesonychians" were dominant instead, representing a very high diversity of eutherian carnivores in the northern continents and Africa. In South America, sparassodonts were dominant instead, while Australia saw the presence of several marsupial predators, such as the dasyuromorphs and thylacoleonids.

In the Mesozoic, while theropod dinosaurs were the larger carnivores, several carnivorous mammal groups were already present. Most notable are the gobiconodontids, the triconodontid *Jugulator*, the deltatheroideans and *Cimolestes*. Many of these, such as *Repenomamus*, *Jugulator* and *Cimolestes*, were among the largest mammals in their faunal assemblages, capable of attacking dinosaurs.

Most carnivorous mammals, from dogs to *Deltatheridium*, share several adaptations in common, such as carnassialiforme teeth, long canines and even similar tooth replacement patterns. Most aberrant are thylacoleonids, which bear a diprodontan dentition completely unlike that of any mammal, and "eutriconodonts" like gobioconodontids and *Jugulator*, by virtue of their cusp anatomy, though they still worked in the same way as carnassials.

Some theropod dinosaurs such as *Tyrannosaurus rex* that existed during the Mesozoic Era were probably obligate carnivores.

Herbivore

A herbivore is an animal anatomically and physiologically adapted to eating plant material, for example foliage, for the main component of its diet. As a result of their plant diet, herbivorous animals typically have mouthparts adapted to rasping or grinding. Horses and other herbivores have wide flat teeth that are adapted to grinding grass, tree bark, and other tough plant material.

A deer and two fawns feeding on foliage

A caterpillar feeding on a leaf

A large percentage of herbivores have mutualistic gut flora that help them digest plant matter, which is more difficult to digest than animal prey. This gut flora is made up of cellulose-digesting protozoans or bacteria living in the herbivores' intestines.

Etymology

Herbivore is the anglicized form of a modern Latin coinage, *herbivora,* cited in Charles Lyell's 1830 *Principles of Geology.* Richard Owen employed the anglicized term in an 1854 work on fossil teeth and skeletons. *Herbivora* is derived from the Latin *herba* meaning a small plant or herb, and *vora,* from *vorare,* to eat or devour.

Definition and Related Terms

Herbivory is a form of consumption in which an organism principally eats autotrophs such as plants, algae and photosynthesizing bacteria. More generally, organisms that feed on autotrophs in general are known as primary consumers. *Herbivory* usually refers to animals eating plants; fungi, bacteria and protists that feed on living plants are usually termed plant pathogens (plant diseases), and microbes that feed on dead plants are saprotrophs. Flowering plants that obtain nutrition from other living plants are usually termed parasitic plants. There is, however, no single exclusive and definitive ecological classification of consumption patterns; each textbook has its own variations on the theme.

Evolution of Herbivory

A fossil *Viburnum lesquereuxii* leaf with evidence of insect herbivory; Dakota Sandstone (Cretaceous) of Ellsworth County, Kansas. Scale bar is 10 mm.

Our understanding of herbivory in geological time comes from three sources; fossilized plants, which may preserve evidence of defence (such as spines), or herbivory-related damage; the observation of plant debris in fossilised animal faeces; and the construction of herbivore mouthparts.

Although herbivory was long thought to be a Mesozoic phenomenon, fossils have shown that within less than 20 million years after the first land plants evolved, plants were being consumed by arthropods. Insects fed on the spores of early Devonian plants, and the Rhynie chert also provides evidence that organisms fed on plants using a "pierce and suck" technique.

During the next 75 million years, plants evolved a range of more complex organs, such as roots and seeds. There is no evidence of any organism being fed upon until the middle-late Mississippian, 330.9 million years ago. There was a gap of 50 to 100 million years between the time each organ evolved and the time organisms evolved to feed upon them; this may be due to the low levels of oxygen during this period, which may have suppressed evolution. Further than their arthropod status, the identity of these early herbivores is uncertain. Hole feeding and skeletonisation are recorded in the early Permian, with surface fluid feeding evolving by the end of that period.

Herbivory among four-limbed terrestrial vertebrates, the tetrapods developed in the Late Carboniferous (307 - 299 million years ago). Early tetrapods were large amphibious piscivores. While amphibians continued to feed on fish and insects, some reptiles began exploring two new food types, tetrapods (carnivory) and plants (herbivory). The entire dinosaur order ornithischia was composed with herbivores dinosaurs. Carnivory was a natural transition from insectivory for medium and large tetrapods, requiring minimal adaptation. In contrast, a complex set of adaptations was necessary for feeding on highly fibrous plant materials.

Arthropods evolved herbivory in four phases, changing their approach to it in response to changing plant communities. Tetrapod herbivores made their first appearance in the fossil record of their jaws near the Permio-Carboniferous boundary, approximately 300 million years ago. The earliest evidence of their herbivory has been attributed to dental occlusion, the process in which teeth from the upper jaw come in contact with teeth in the lower jaw is present. The evolution of dental occlusion led to a drastic increase in plant food processing and provides evidence about feeding strategies based on tooth wear patterns. Examination of phylogenetic frameworks of tooth and jaw morphologes has revealed that dental occlusion developed independently in several lineages tetrapod herbivores. This suggests that evolution and spread occurred simultaneously within various lineages.

Food chain

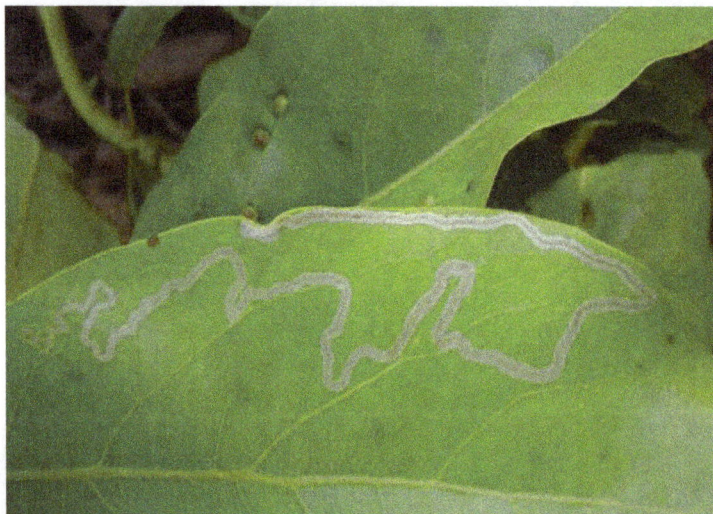

Leaf miners feed on leaf tissue between the epidermal layers, leaving visible trails

Herbivores form an important link in the food chain; because they consume plants in order to digest the carbohydrates photosynthetically produced by a plant. Carnivores in turn consume herbivores for the same reason, while omnivores can obtain their nutrients from either plants or animals. Due to a herbivore's ability to survive solely on tough and fibrous plant matter, they are termed the primary consumers in the food cycle (chain). Herbivory, carnivory, and omnivory can be regarded as special cases of Consumer-Resource Systems.

Feeding Strategies

Two herbivore feeding strategies are grazing (e.g. cows) and browsing (e.g. moose). Although the exact definition of the feeding strategy may depend on the writer, most authors agree that to define a grazer at least 90% of the forage has to be grass, and for a browser at least 90% tree leaves and/or twigs. An intermediate feeding strategy is called "mixed-feeding". In their daily need to take up energy from forage, herbivores of different body mass may be selective in choosing their food. "Selective" means that herbivores may choose their forage source depending on, e.g., season or food availability, but also that they may choose high quality (and consequently highly nutritious) forage before lower quality. The latter especially is determined by the body mass of the herbivore, with small herbivores selecting for high quality forage, and with increasing body mass animals are less selective. Several theories attempt to explain and quantify the relationship between animals and their food, such as Kleiber's law, Holling's disk equation and the marginal value theorem.

Kleiber's law describes the relationship between an animal's size and its feeding strategy, saying that larger animals need to eat less food per unit weight than smaller animals. Kleiber's law states that the metabolic rate (q_0) of an animal is the mass of the animal (M) raised to the 3/4 power: $q_0 = M^{3/4}$ Therefore, the mass of the animal increases at a faster rate than the metabolic rate.

Herbivores employ numerous types of feeding strategies. Many herbivores do not fall into one specific feeding strategy, but employ several strategies and eat a variety of plant parts.

Types of feeding strategies		
Feeding Strategy	**Diet**	**Example**
Algivores	Algae	krill, crabs, sea snail, sea urchin, parrotfish, surgeonfish, flamingo
Frugivores	Fruit	Ruffed lemurs
Folivores	Leaves	Koalas
Nectarivores	Nectar	Honey possum
Granivores	Seeds	Hawaiian honeycreepers
Palynivores	Pollen	Bees
Mucivores	Plant fluids, i.e. sap	Aphids
Xylophages	Wood	Termites

Optimal Foraging Theory is a model for predicting animal behavior while looking for food or other resource, such as shelter or water. This model assesses both individual movement, such as animal behavior while looking for food, and distribution within a habitat, such as dynamics at the population and community level. For example, the model would be used to look at the browsing behavior of a deer while looking for food, as well as that deer's specific location and movement within the forested habitat and its interaction with other deer while in that habitat.

This model has been criticized as circular and untestable. Critics have pointed out that its proponents use examples that fit the theory, but do not use the model when it does not fit the reality. Other critics point out that animals do not have the ability to assess and maximize their potential gains, therefore the optimal foraging theory is irrelevant and derived to explain trends that do not exist in nature.

Holling's disk equation models the efficiency at which predators consume prey. The model predicts that as the number of prey increases, the amount of time predators spend handling prey also increases and therefore the efficiency of the predator decreases.In 1959, S. Holling proposed an equation to model the rate of return for an optimal diet: Rate (R) = Energy gained in foraging (Ef)/ (time searching (Ts) + time handling (Th))

$$R = Ef / (Ts + Th)$$

Where s = cost of search per unit time f = rate of encounter with items, h = handling time, e = energy gained per encounterIn effect, this would indicate that a herbivore in a dense forest would spend more time handling (eating) the vegetation because there was so much vegetation around than a herbivore in a sparse forest, who could easily browse through the forest vegetation. According to the Holling's disk equation, a herbivore in the sparse forest would be more efficient at eating than the herbivore in the dense forest.

The marginal value theorem describes the balance between eating all the food in a patch for immediate energy, or moving to a new patch and leaving the plants in the first patch to regenerate for future use. The theory predicts that absent complicating factors, an animal should leave a resource patch when the rate of payoff (amount of food) falls below the average rate of payoff for the entire area. According to this theory, locus should move to a new patch of food when the patch they are currently feeding on requires more energy to obtain food than an average patch. Within this theory, two subsequent parameters emerge, the Giving Up Density (GUD) and the Giving Up Time (GUT). The Giving Up Density (GUD) quantifies the amount of food that remains in a patch when a forager moves to a new patch. The Giving Up Time (GUT) is used when an animal continuously assesses the patch quality.

Attacks and Counter-attacks

Herbivore Offense

Aphids are fluid feeders on plant sap.

The myriad defenses displayed by plants means that their herbivores need a variety of skills to overcome these defenses and obtain food. These allow herbivores to increase their feeding and use of a host plant. Herbivores have three primary strategies for dealing with plant defenses; choice, herbivore modification, and plant modification.

Feeding choice involves which plants a herbivore chooses to consume. It has been suggested that many herbivores feed on a variety of plants to balance their nutrient uptake and to avoid consuming too much of any one type of defensive chemical. This involves a tradeoff however, between foraging on many plant species to avoid toxins or specializing on one type of plant that can be detoxified.

Herbivore modification is when various adaptations to body or digestive systems of the herbivore allow them to overcome plant defenses. This might include detoxifying secondary metabolites, sequestering toxins unaltered, or avoiding toxins, such as through the production of large amounts of saliva to reduce effectiveness of defenses. Herbivores may also utilize symbionts to evade plant defences. For example, some aphids use bacteria in their gut to provide essential amino acids lacking in their sap diet.

Plant modification occurs when herbivores manipulate their plant prey to increase feeding. For example, some caterpillars roll leaves to reduce the effectiveness of plant defenses activated by sunlight.

Plant Defense

A plant defense is a trait that increases plant fitness when faced with herbivory. This is measured relative to another plant that lacks the defensive trait. Plant defenses increase survival and/or reproduction (fitness) of plants under pressure of predation from herbivores.

Defense can be divided into two main categories, tolerance and resistance. Tolerance is the ability of a plant to withstand damage without a reduction in fitness. This can occur by diverting herbivory to non-essential plant parts or by rapid regrowth and recovery from herbivory. Resistance refers to the ability of a plant to reduce the amount of damage it receives from a herbivore. This can occur via avoidance in space or time, physical defenses, or chemical defenses. Defenses can either be constitutive, always present in the plant, or induced, produced or translocated by the plant following damage or stress.

Physical, or mechanical, defenses are barriers or structures designed to deter herbivores or reduce intake rates, lowering overall herbivory. Thorns such as those found on roses or acacia trees are one example, as are the spines on a cactus. Smaller hairs known as trichomes may cover leaves or stems and are especially effective against invertebrate herbivores. In addition, some plants have waxes or resins that alter their texture, making them difficult to eat. Also the incorporation of silica into cell walls is analogous to that of the role of lignin in that it is a compression-resistant structural component of cell walls; so that plants with their cell walls impregnated with silica are thereby afforded a measure of protection against herbivory.

Chemical defenses are secondary metabolites produced by the plant that deter herbivory. There are a wide variety of these in nature and a single plant can have hundreds of different chemical defenses. Chemical defenses can be divided into two main groups, carbon-based defenses and nitrogen-based defenses.

1. Carbon-based defenses include terpenes and phenolics. Terpenes are derived from 5-carbon isoprene units and comprise essential oils, carotenoids, resins, and latex. They can have a number of functions that disrupt herbivores such as inhibiting adenosine triphosphate (ATP) formation, molting hormones, or the nervous system. Phenolics combine an aromatic carbon ring with a hydroxyl group. There are a number of different phenolics such as lignins, which are found in cell walls and are very indigestible except for specialized microorganisms; tannins, which have a bitter taste and bind to proteins making them indigestible; and furanocumerins, which produce free radicals disrupting DNA, protein, and lipids, and can cause skin irritation.

2. Nitrogen-based defenses are synthesized from amino acids and primarily come in the form of alkaloids and cyanogens. Alkaloids include commonly recognized substances such as caffeine, nicotine, and morphine. These compounds are often bitter and can inhibit DNA or RNA synthesis or block nervous system signal transmission. Cyanogens get their name from the cyanide stored within their tissues. This is released when the plant is damaged and inhibits cellular respiration and electron transport.

Plants have also changed features that enhance the probability of attracting natural enemies to herbivores. Some emit semiochemicals, odors that attract natural enemies, while others provide food and housing to maintain the natural enemies' presence, e.g. ants that reduce herbivory. A given plant species often has many types of defensive mechanisms, mechanical or chemical, constitutive or induced, which allow it to escape from herbivores.

Herbivore–plant Interactions Per Predator–prey Theory

According to the theory of predator–prey interactions, the relationship between herbivores and plants is cyclic. When prey (plants) are numerous their predators (herbivores) increase in numbers, reducing the prey population, which in turn causes predator number to decline. The prey population eventually recovers, starting a new cycle. This suggests that the population of the herbivore fluctuates around the carrying capacity of the food source, in this case the plant.

Several factors play into these fluctuating populations and help stabilize predator–prey dynamics. For example, spatial heterogeneity is maintained, which means there will always be pockets of plants not found by herbivores. This stabilizing dynamic plays an especially important role for specialist herbivores that feed on one species of plant and prevents these specialists from wiping out their food source. Prey defenses also help stabilize predator–prey dynamics. Eating a second prey type helps herbivores' populations stabilize. Alternating between two or more plant types provides population stability for the herbivore, while the populations of the plants oscillate. This plays an important role for generalist herbivores that eat a variety of plants. Keystone herbivores keep vegetation populations in check and allow for a greater diversity of both herbivores and plants. When an invasive herbivore or plant enters the system, the balance is thrown off and the diversity can collapse to a monotaxon system.

The back and forth relationship of plant defense and herbivore offense can be seen as a sort of "adaptation dance" in which one partner makes a move and the other counters it. This reciprocal change drives coevolution between many plants and herbivores, resulting in what has been referred to as a "coevolutionary arms race". The escape and radiation mechanisms for coevolution, presents the idea that adaptations in herbivores and their host plants, has been the driving force behind speciation.

While much of the interaction of herbivory and plant defense is negative, with one individual reducing the fitness of the other, some is actually beneficial. This beneficial herbivory takes the form of mutualisms in which both partners benefit in some way from the interaction. Seed dispersal by herbivores and pollination are two forms of mutualistic herbivory in which the herbivore receives a food resource and the plant is aided in reproduction.

Impacts

Herbivorous fish and marine animals are an indispensable part of the coral reef ecosystem. Since algae and seaweeds grow much faster than corals they can occupy spaces where corals could have settled. They can outgrow and thus outcompete corals on bare surfaces. In the absence of plant-eating fish, seaweeds deprive corals of sunlight. They can also physically damage corals with scrapes.

The impact of herbivory can be seen in areas ranging from economics to ecological, and both. For example, environmental degradation from white-tailed deer (Odocoileus virginianus) in the US alone has the potential to both change vegetative communities through over-browsing and cost forest restoration projects upwards of $750 million annually. Agricultural crop damage by the same species totals approximately $100 million every year. Insect crop damages also contribute largely to annual crop losses in the U.S. Herbivores affect economics through the revenue generated by hunting and ecotourism. For example, the hunting of herbivorous game species such as white-tailed deer, cottontail rabbits, antelope, and elk in the U.S. contributes greatly to the billion-dollar annually hunting industry. Ecotourism is a major source of revenue, particularly in Africa, where many large mammalian herbivores such as elephants, zebras, and giraffes help to bring in the equivalent of millions of US dollars to various nations annually.

Omnivore

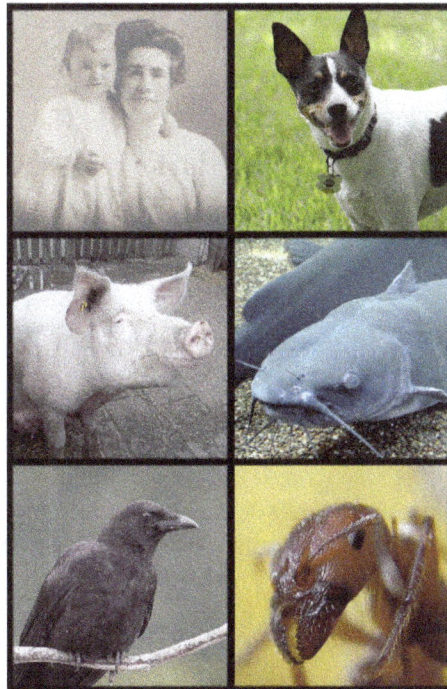

Examples of omnivores. From left to right: Humans, dogs, pigs, walking catfish, American crow, gravel ant.

Omnivore is a consumption classification for animals that have the capability to obtain chemical energy and nutrients from materials originating from plant and animal origin. Often, omnivores also have the ability to incorporate food sources such as algae, fungi, and bacteria into their diet as well.

Omnivores come from diverse backgrounds that often independently evolved sophisticated consumption capabilities. For instance, dogs evolved from primarily carnivorous organisms (Carnivora) while pigs evolved from primarily herbivorous organisms (Artiodactyla). What this means is that physical characteristics are often not reliable indicators of whether an animal has the ability to obtain energy and nutrients from both plant and animal matter. Due to the wide range of entirely unrelated organisms independently evolving the capability to obtain energy and nutrients from both plant and animal materials, no generalizations about the anatomical features of all omnivores can realistically be made.

The variety of different animals that are classified as omnivores can be placed into further categories depending on their feeding behaviors. Frugivores include maned wolves and orangutans; insectivores include swallows and pink fairy armadillos; granivores include large ground finches and humans. (This is due to the average human diet mainly consisting of grains, with rice, maize and wheat comprising two-thirds of human food consumption).

All of these animals are omnivores, yet still fall into special niches in terms of feeding behavior and preferred foods. Being omnivores gives these animals more food security in stressful times or makes possible living in less consistent environments.

Etymology and Definitions

The word *omnivore* derives from the Latin *omnis* (all), and *vora*, from vorare, (to eat or devour), having been coined by the French and later adopted by the English in the 1800s. Traditionally the definition for omnivory was entirely behavioral by means of simply "including both animal and vegetable tissue in the diet." In more recent times, with the advent of advanced technological capabilities in fields like gastroenterology, biologists have formulated a standardized variation of omnivore used for labeling a species actual ability to obtain energy and nutrients from materials. This has subsequently conditioned two context specific definitions.

- Behavioral: This definition is used to specify if a species or individual is actively consuming both plant and animal materials. (e.g. "vegans do not participate in the omnivore based diet.")

- Physiological: This definition is often used in academia to specify species that have the capability to obtain energy and nutrients from both plant and animal matter. (e.g. "humans are omnivores due to their capability to obtain energy and nutrients from both plant and animal materials.")

The taxonomic utility of omnivore's traditional and behavioral definition is limited, since the diet, behavior, and phylogeny of one omnivorous species might be very different from that of another: for instance, an omnivorous pig digging for roots and scavenging for fruit and carrion is taxonomically and ecologically quite distinct from an omnivorous chameleon that eats leaves and insects. The term "omnivory" is also not always comprehensive because it does not deal with mineral foods such as salt licks and the consumption of plant and animal material for medical purposes which would not otherwise be consumed (i.e. zoopharmacognosy) within non-omnivores.

Classification, Contradictions and Difficulties

Though Carnivora is a taxon for species classification, no such equivalent exists for omnivores, as omnivores are widespread across multiple taxonomic clades. The Carnivora order does not include all carnivorous species, and not all species within the Carnivora taxon are carnivorous. It is common to find physiological carnivores consuming materials from plants or physiological herbivores consuming material from animals, e.g. felines eating grass and deer eating birds. From a behavioral aspect, this would make them omnivores, but from the physiological standpoint, this may be due to zoopharmacognosy. Physiologically, animals must be able to obtain both energy and nutrients from plant and animal materials to be considered omnivorous. Thus, such animals are still able to be classified as carnivores and herbivores when they are just obtaining nutrients from materials originating from sources that do not seemingly complement their classification. For instance, it is well documented that animals such as giraffes, camels, and cattle will gnaw on bones, preferably dry bones, for particular minerals and nutrients. Felines, which are usually regarded as obligate carnivores, occasionally eat grass to regurgitate indigestibles (e.g. hair, bones), aid with hemoglobin production, and as a laxative.

Occasionally, it is found that animals historically classified as carnivorous may deliberately eat plant material. For example, in 2013 it was considered that American alligators (*Alligator mississippiensis*) may be physiologically omnivorous once investigations had been conducted on why they occasionally eat fruits. It was suggested that alligators probably ate fruits both accidentally but also deliberately.

"Life-history omnivores" is a specialized classification given to organisms that change their eating habits during their life cycle. Some species, such as grazing waterfowl like geese, are known to eat mainly animal tissue at one stage of their lives, but plant matter at another. The same is true for many insects, such as beetles in the family Meloidae, which begin by eating animal tissue as larvae, but change to eating plant matter after they mature. Likewise, many mosquito species in early life eat plants or assorted detritus, but as they mature, males continue to eat plant matter and nectar whereas the females (such as those of *Anopheles*, *Aedes* and *Culex*) also eat blood to reproduce effectively.

Omnivorous Species

General

Although cases exist of carnivores eating plant matter and herbivores eating meat, the classification "omnivore" refers to the adaptations and main food source of the species in general, so these exceptions do not make either individual animals or the species as a whole omnivorous. For the concept of "omnivore" to be regarded as a scientific classification, some clear set of measurable and relevant criteria would need to be considered to differentiate between an "omnivore" and other categories, e.g. faunivore, folivore, and scavenger. Some researchers argue that evolution of any species from herbivory to carnivory or carnivory to herbivory would be rare except via an intermediate stage of omnivory.

Omnivorous Mammals

Various mammals are omnivorous in the wild, such as species of pigs, badgers, bears, coatis, civets, hedgehogs, opossums, skunks, sloths, squirrels, raccoons, chipmunks, mice, and rats. Hominidae, including humans and chimpanzees, are also omnivores.

Most bear species are omnivores, but individual diets can range from almost exclusively herbivorous to almost exclusively carnivorous, depending on what food sources are available locally and seasonally. Polar bears are classified as carnivores, both taxonomically (they are in the order Carnivora), and behaviorally (they subsist on a largely carnivorous diet). Depending on the species of bear, there is generally a preference for one class of food, as plants and animals are digested differently. Wolf subspecies (including wolves, dogs, dingoes, and coyotes) have a general preference and are evolutionarily geared towards meat, but also will voluntarily eat plant material like fruits, vegetables, and grasses, and can live on such indefinitely. Also, the maned wolf is a canid whose diet is naturally 50% plant matter.

Most bear species are omnivores

While most mammals may display "omnivorous" behavior patterns depending on conditions of supply, culture, season and so on, they will generally prefer a particular class of food, to which their digestive processes are adapted. Like most arboreal species, most squirrels are primarily granivores, subsisting on nuts and seeds. But like virtually all mammals, squirrels avidly consume some animal food when it becomes available. For example, the American eastern gray squirrel has been introduced by humans to parts of Britain, continental Europe and South Africa. Where it flourishes, its effect on populations of nesting birds is often serious, largely because of consumption of eggs and nestlings.

Other Species

Various birds are omnivorous, with diets varying from berries and nectar to insects, worms, fish, and small rodents. Examples include cassowaries, chickens, crows and related corvids, keas, rallidae, and rheas. In addition, some lizards, turtles, fish (such as piranhas and catfish), and invertebrates are also omnivorous.

Quite often, mainly herbivorous creatures will eagerly eat small quantities of animal food when it becomes available. Although this is trivial most of the time, omnivorous or herbivorous birds, such as sparrows, often will feed their chicks insects while food is most needed for growth. On close inspection it appears that nectar-feeding birds such as sunbirds rely on the ants and other insects that they find in flowers, not for a richer supply of protein, but for essential nutrients such as cyanocobalamin that are absent from nectar. Similarly, monkeys of many species eat maggoty fruit, sometimes in clear preference to sound fruit. When to refer to such animals as omnivorous, or otherwise, is a question of context and emphasis, rather than of definition.

Parasitism

In biology/ecology, parasitism is a non-mutual relationship between species, where one species, the parasite, benefits at the expense of the other, the host. Traditionally *parasite* (in biological usage) referred primarily to organisms visible to the naked eye, or macroparasites (such as helminths). Parasites can be microparasites, which are typically smaller, such as protozoa, viruses, and bacteria. Examples of parasites include the plants mistletoe and cuscuta, and animals such as hookworms.

A *Lithognathus* fish parasitised by a *Cymothoa exigua* parasite

Unlike predators, parasites typically do not kill their host, are generally much smaller than their host, and will often live in or on their host for an extended period. Both are special cases of consumer-resource interactions. Parasites show a high degree of specialization, and reproduce at a faster rate than their hosts. Classic examples of parasitism include interactions between vertebrate hosts and tapeworms, flukes, the *Plasmodium* species, and fleas. Parasitoidy is an evolutionary strategy within parasitism in which the parasite generally kills its host.

In brood parasitism, the host raises the young of another species, here the egg of a cowbird, that has been laid in its nest.

Parasites reduce host biological fitness by general or specialized pathology, such as parasitic castration and impairment of secondary sex characteristics, to the modification of host behavior. Parasites increase their own fitness by exploiting hosts for resources necessary for their survival, such as food, water, heat, habitat, and transmission. Although parasitism applies unambiguously to many cases, it is part of a continuum of types of interactions between species, rather than an exclusive category. In many cases, it is difficult to demonstrate harm to the host. In others, there may be no apparent specialization on the part of the parasite, or the interaction between the organisms may remain short-lived.

Etymology

First used in English 1539, the word *parasite* comes from the Medieval French *parasite*, from the Latin *parasitus*, the latinisation of the Greek (*parasitos*), "one who eats at the table of another" and that from (*para*), "beside, by" + (*sitos*), "wheat". Coined in English in 1611, the word *parasitism* comes from the Greek (*para*) + (*sitismos*) "feeding, fattening". In its original sense, it was not strictly pejorative in nature; being a *parasitos* was an accepted lifestyle, whereby a person could live off the hospitality of others, and in return provide "flattery, simple services, and a willingness to endure humiliation".

Types

Parasites are classified based on their interactions with their hosts and on their life cycles. An obligate parasite is totally dependent on the host to complete its life cycle, while a facultative parasite is not. A direct parasite has only one host while an indirect parasite has multiple hosts. For indirect parasites, there will always be a definitive host and an intermediate host.

Acrodactyla quadrisculpta is a wasp parasitoid of spiders. Parasitoids are parasites that eventually kill their hosts.

There are six major evolutionary strategies within parasitism, namely parasitic castrator, directly transmitted parasite, trophically transmitted parasite, vector-transmitted parasite, parasitoid (which eventually kills the host), and micropredator. These strategies for successful parasitism are adaptive peaks; many intermediate strategies are possible, but organisms in many different groups have consistently converged on these six, which are evolutionarily stable.

Human head lice (*Pediculus humanus capitis*) are obligate ectoparasites.

Parasites that live on the outside of the host, either on the skin or the outgrowths of the skin, are called ectoparasites (e.g. lice, fleas, and some mites).

Those that live inside the host are called endoparasites (including all parasitic worms). Endoparasites can exist in one of two forms: intercellular parasites (inhabiting spaces in the host's body) or intracellular parasites (inhabiting cells in the host's body). Intracellular parasites, such as protozoa, bacteria or viruses, tend to rely on a third organism, which is generally known as the carrier or vector. The vector does the job of transmitting them to the host. An example of this interaction is the transmission of malaria, caused by a protozoan of the genus *Plasmodium*, to humans by the bite of an anopheline mosquito.

Those parasites living in an intermediate position, being half-ectoparasites and half-endoparasites, are called mesoparasites.

An epiparasite is one that feeds on another parasite. This relationship is also sometimes referred to as *hyperparasitism,* exemplified by a protozoan (the hyperparasite) living in the digestive tract of a flea living on a dog.

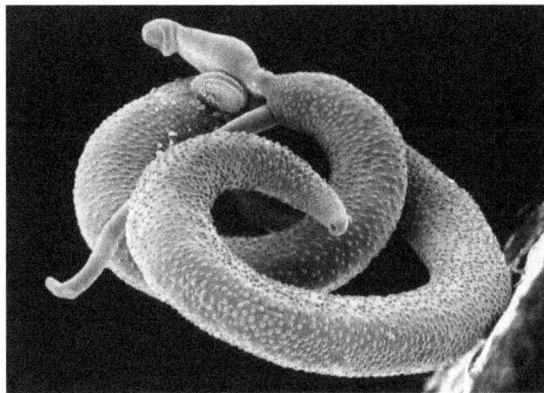

Schistosoma mansoni is an obligate endoparasite of human blood vessels.

Social parasites take advantage of interactions between members of social organisms such as ants, termites, and bumblebees. Examples include *Phengaris arion*, a butterfly whose larvae employ mimicry to parasitize certain species of ants, *Bombus bohemicus*, a bumblebee who invades the hives of other species of bee and takes over reproduction, their young raised by host workers, and *Melipona scutellaris,* a eusocial bee where virgin queens escape killer workers and invade another colony without a queen. An extreme example of social parasitism is the ant species of *Tetramorium inquilinum* of the Alps, which spend their whole lives on the back of *Tetramorium* host ants. With tiny and deprecated bodies they have evolved for one single task: holding on to their host. If they fall off, they most likely would not have the strength to climb back on top of another ant, and eventually they will die.

In kleptoparasitism (from the Greek (kleptes), thief), parasites appropriate food gathered by the host. An example is the brood parasitism practiced by cowbirds, whydahs, cuckoos, and black-headed ducks which do not build nests of their own and leave their eggs in nests of other species. The host behaves as a "babysitter" as they raise the young as their own. If the host removes the cuckoo's eggs, some cuckoos will return and attack the nest to compel host birds to remain subject to this parasitism.

Intraspecific social parasitism may also occur. One example of this is *parasitic nursing*, where some individuals take milk from unrelated females. In wedge-capped capuchins, higher ranking females sometimes take milk from low ranking females without any reciprocation. The high ranking females benefit at the expense of the low ranking females.

Parasitism can take the form of isolated *cheating* or *exploitation* among more generalized mutualistic interactions. For example, broad classes of plants and fungi exchange carbon and nutrients in common mutualistic mycorrhizal relationships; however, some plant species known as myco-heterotrophs "cheat" by taking carbon from a fungus rather than donating it.

An adelpho-parasite (from the Greek (adelphos), brother) is a parasite in which the host species is closely related to the parasite, often being a member of the same family or genus. An example of this is the citrus blackfly parasitoid, *Encarsia perplexa*, unmated females of which may lay haploid eggs in the fully developed larvae of their own species. These result in the production of male offspring. The marine worm *Bonellia viridis* has a similar reproductive strategy, although the larvae are planktonic.

Autoinfection is the infection of a primary host with a parasite, particularly a helminth, in such a way that the complete life cycle of the parasite happens in a single organism, without the involvement of another host. Therefore, the primary host is at the same time the secondary host of the parasite. Some of the organisms where autoinfection occurs are *Strongyloides stercoralis*, *Enterobius vermicularis*, *Taenia solium*, and *Hymenolepis nana*. Strongyloidiasis for example involves premature transformation of noninfective larvae in infective larvae, which can then penetrate the intestinal mucosa (internal autoinfection) or the skin of the perineal area (external autoinfection). Infection can be maintained by repeated migratory cycles for the remainder of the person's life.

Host Defenses

In Vertebrates

The first line of defense against invading parasites in vertebrates is the skin. Skin is made up of layers of dead cells and acts as a physical barrier to invading organisms. These dead cells contain the protein keratin, which makes skin tough and waterproof. Most microorganisms need a moist environment to survive. By keeping the skin dry, it prevents invading organisms from colonizing. Furthermore, human skin also secretes sebum, which is toxic to most microorganisms.

The vertebrate mouth contains saliva, which prevents foreign organisms from getting into the body orally. Furthermore, the mouth also contains lysozyme, an enzyme found in tears and the saliva. This enzyme breaks down cell walls of invading microorganisms.

Should the organism pass the mouth, the stomach is the next line of defense. The vertebrate stomach contains hydrochloric acid and gastric acids, which makes its pH level around 2. In this environment, the acidity of the stomach helps kill most microorganisms that try to invade the body through the gastric intestinal tract.

Parasites can also invade the body through the eyes. The lashes on the eyelids of mammals prevents invading microorganisms from entering the eye in the first place. Even if the microorganism does get into the eye, tears contain the enzyme lysozyme, which will kill most invading microorganisms.

Should the parasite enter the body, the immune system is a vertebrate's major defense against parasitic invasion. The immune system is made up of different families of molecules. These include serum proteins and pattern recognition receptors (PRRs). PRRs are intracellular and cellular receptors that activate dendritic cells, which in turn activate the adaptive immune system's lymphocytes. Lymphocytes such as the T cells and antibody producing B cells with variable receptors that recognize parasites.

In Insects

Insects often adapt their nests to aid in parasite defense. For example, one of the key reasons the *Polistes canadensis* nests across multiple combs rather than building a single comb like much of the rest of its genus is as a defense mechanism against the infestation of tineid moths. The tineid moth lays its eggs within the wasps' nests and then these eggs hatch into larvae that can burrow from cell to cell and prey on wasp pupae. Adult wasps attempt to remove and kill moth eggs and larvae by chewing down the edges of cells, coating the cells with an oral secretion that gives the nest a dark brownish appearance.

In Plants

In response to parasitic attack, plants undergo a series of metabolic and biochemical reaction pathways that will enact defensive responses. For example, parasitic invasion causes an increase in the jasmonic acid-insensitivel (JA) and NahG (SA) pathway. These pathways produce chemicals that induce defensive responses, such as the production of chemicals or defensive molecules to fight off the attack. Different biochemical pathways are activated by different parasites. In general, there are two types of responses that can be activated by the pathways. Plants can either initiate a specific or non-specific response. Specific responses involve gene-gene recognition of the plant and parasite. This can be mediated by the ability of the plant's cell receptors recognizing and binding molecules that are located on the cell surface of parasites. Once the plant's receptors recognizes the parasite, the plant localizes the defensive compounds to that area creating a hypersensitive response. This form of defense mechanism localizes the area of attack and keeps the parasite from spreading. Furthermore, a specific response against parasitic attack prevents the plants from wasting its energy by increasing defenses where it's not needed. However, specific defensive responses only target specific parasites. If the plant lacks the ability to recognize a parasite, specific defense responses won't be activated. Nonspecific defensive responses work against all parasites. These responses are active over time and are systematic, meaning that the responses are not confined to an area of the plant, but rather spread throughout the entirety of the organism. However, nonspecific responses are energy costly, since the plant has to ensure that the genes producing the nonspecific responses are always expressed.

Evolutionary Aspects

Parasitism has arisen independently many times. Depending on the definition used, as many as half of all animals have at least one parasitic phase in their life cycles, and it is frequent in plants and fungi. Almost all free-living animals are host to one or more parasitic taxa.

Parasites evolve in response to their hosts' defences, sometimes in a manner specific to a particular host taxon and specializing to the point where they infect only a single species. Such narrow host specificity can be costly over evolutionary time, however, if the host species becomes extinct. Therefore, many parasites can infect a variety of more or less closely related host species, with different success rates.

Restoration of a *Tyrannosaurus* with parasite infections. A 2009 study showed that holes in the skulls of several specimens might have been caused by *Trichomonas*-like parasites

In turn, host defenses coevolve in response to attacks by parasites. Theoretically, parasites may have an advantage in this evolutionary arms race because their generation time commonly is shorter. Hosts reproduce less quickly than parasites, and therefore have fewer chances to adapt than their parasites do over a given span of time.

Long-term coevolution sometimes leads to a relatively stable relationship tending to commensalism or mutualism, as, all else being equal, it is in the evolutionary interest of the parasite that its host thrives. A parasite may evolve to become less harmful for its host or a host may evolve to cope with the unavoidable presence of a parasite—to the point that the parasite's absence causes the host harm. For example, although animals infected with parasitic worms are often clearly harmed, and therefore parasitized, such infections may also reduce the prevalence and effects of autoimmune disorders in animal hosts, including humans. In a more extreme example, some nematode worms cannot reproduce, or even survive, without infection by Wolbachia bacteria.

Competition between parasites tends to favor faster reproducing and therefore more virulent parasites. Parasites whose life cycle involves the death of the host, to exit the present host and sometimes to enter the next, evolve to be more virulent or even alter the behavior or other properties of the host to make it more vulnerable to predators. Parasites whose reproduction is largely tied their hosts' reproductive success tend to become less virulent or mutualist, so that its hosts reproduce more effectively.

The presumption of a shared evolutionary history between parasites and hosts can sometimes elucidate how host taxa are related. For instance, there has been dispute about whether flamingos are more closely related to the storks and their relatives, or to ducks, geese and their relatives. The fact that flamingos share parasites with ducks and geese is evidence these groups may be more closely related to each other than either is to storks.

Parasitism is part of one explanation for the evolution of secondary sex characteristics seen in breeding males throughout the animal world, such as the plumage of male peacocks and manes of male lions. According to this theory, female hosts select males for breeding based on such characteristics because they indicate resistance to parasites and other disease.

Co-speciation

In rare cases, a parasite may even undergo co-speciation with its host. One particularly remarkable example of co-speciation exists between the simian foamy virus (SFV) and its primate hosts. In one study, the phylogenies of SFV polymerase and the mitochondrial cytochrome oxidase subunit II from African and Asian primates were compared. Surprisingly, the phylogenetic trees were very congruent in branching order and divergence times. Thus, the simian foamy viruses may have co-speciated with Old World primates for at least 30 million years.

Evolutionary events like host switch, host shift, the duplication or extinction of parasite species (without similar events on the host phylogeny) often erode topographical similarities between host and parasite phylogenies.

Ecology

Quantitative Ecology

A single parasite species usually has an aggregated distribution across host individuals, which means that most hosts harbor few parasites, while a few hosts carry the vast majority of parasite individuals. This poses considerable problems for students of parasite ecology: the use of parametric statistics should be avoided. Log-transformation of data before the application of parametric test, or the use of non-parametric statistics is recommended by several authors. However, this can give rise to further problems. Therefore, modern day quantitative parasitology is based on more advanced biostatistical methods.

Diversity Ecology

Hosts represent discrete habitat patches that can be occupied by parasites. A hierarchical set of terminology has come into use to describe parasite assemblages at different host scales.

Infrapopulation

All the parasites of one species in a single individual host.

Metapopulation

All the parasites of one species in a host population.

Infracommunity

All the parasites of all species in a single individual host.

Compound Community

All the parasites of all species in all host species in an ecosystem.

The diversity ecology of parasites differs markedly from that of free-living organisms. For free-living organisms, diversity ecology features many strong conceptual frameworks including Robert MacArthur and E. O. Wilson's theory of island biogeography, Jared Diamond's assembly rules and, more recently, null models such as Stephen Hubbell's unified neutral theory of biodiversity and biogeography. Frameworks are not so well-developed for parasites and in many ways they do not fit the free-living models. For example, island biogeography is predicated on fixed spatial relationships between habitat patches ("sinks"), usually with reference to a mainland ("source"). Parasites inhabit hosts, which represent mobile habitat patches with dynamic spatial relationships. There is no true "mainland" other than the sum of hosts (host population), so parasite component communities in host populations are metacommunities.

Nonetheless, different types of parasite assemblages have been recognized in host individuals and populations, and many of the patterns observed for free-living organisms are also pervasive among parasite assemblages. The most prominent of these is the interactive-isolationist continuum. This proposes that parasite assemblages occur along a cline from interactive communities, where niches are saturated and interspecific competition is high, to isolationist communities, where there are many vacant niches and interspecific interaction is not as important as stochastic factors in providing structure to the community. Whether this is so, or whether community patterns simply reflect the sum of underlying species distributions (no real "structure" to the community), has not yet been established.

Adaptation

Parasites infect hosts within their same geographical area (sympatric) more effectively. This phenomenon supports the Red Queen hypothesis, which states that interactions between species, such as host and parasites, lead to constant natural selection for co-adaptation. Parasites track the locally common hosts' phenotypes, therefore the parasites are less infective to allopatric (from different geographical region) hosts.

Experiments published in 2000 discuss the analysis of two different snail populations from two different sources—Lake Ianthe and Lake Poerua in New Zealand. The populations were exposed to two pure parasites (digenetic trematode) taken from the same lakes. In the experiment, the snails were infected by their sympatric parasites, allopatric parasites and mixed sources of parasites. The results suggest that the parasites were more highly effective in infecting their sympatric snails than their allopatric snails. Though the allopatric snails were still infected by the parasites, the infectivity was much less when compared to the sympatric snails. Hence, the parasites were found to have adapted to infecting local populations of snails.

Transmission

Parasites use a variety of methods to infect hosts. For example, the *Acanthamoeba* enters the body when the environment is not hostile, and *Strongyloides stercoralis* enters the body when a host steps on infected ground while barefoot. Many parasites enter the food of their hosts and wait to be eaten. *Plasmodium malariae* uses a mosquito host to transmit malaria, and *Loa loa* parasites use deer flies to enter hosts.

Parasites inhabit living organisms and therefore face problems that free-living organisms do not. Hosts, the only habitats in which parasites can survive, actively try to avoid, repel, and destroy parasites. Parasites employ numerous strategies for getting from one host to another, a process sometimes referred to as parasite *transmission* or *colonization*.

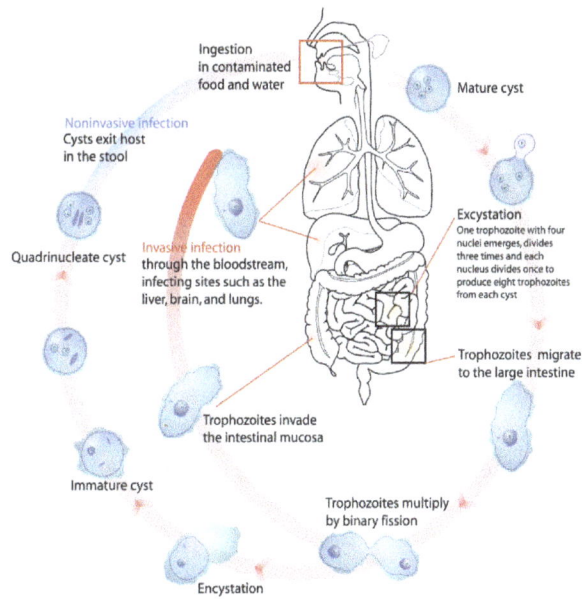

Life cycle of *Entamoeba histolytica*, an anaerobic parasitic protozoan.

Some endoparasites infect their host by penetrating its external surface, while others must be ingested. Once inside the host, adult endoparasites need to shed offspring into the external environment to infect other hosts. Many adult endoparasites reside in the host's gastrointestinal tract, where offspring can be shed along with host excreta. Adult stages of tapeworms, thorny-headed worms and most flukes use this method.

Among protozoan endoparasites, such as the malarial parasites and trypanosomes, infective stages in the host's blood are transported to new hosts by biting-insects, or vectors.

Larval stages of endoparasites often infect sites in the host other than the blood or gastrointestinal tract. In many such cases, larval endoparasites require their host to be consumed by the next host in the parasite's life cycle in order to survive and reproduce. Alternatively, larval endoparasites may shed free-living transmission stages that migrate through the host's tissue into the external environment, where they actively search for or await ingestion by other hosts. The foregoing strategies are used, variously, by larval stages of tapeworms, thorny-headed worms, flukes and parasitic roundworms.

Some ectoparasites, such as monogenean worms, rely on direct contact between hosts. Ectoparasitic arthropods may rely on host-host contact (e.g. many lice), shed eggs that survive off the host (e.g. fleas), or wait in the external environment for an encounter with a host (e.g. ticks). Some aquatic leeches locate hosts by sensing movement and only attach when certain temperature and chemical cues are present.

Some parasites modify host behavior in order to increase the transmission between hosts, often in relation to predator and prey (parasite increased trophic transmission). For example, in California salt marshes, the fluke *Euhaplorchis californiensis* reduces the ability of its killifish host to avoid predators. This parasite matures in egrets, which are more likely to feed on infected killifish than on uninfected fish. Another example is the protozoan *Toxoplasma gondii*, a parasite that matures in cats but can be carried by many other mammals. Uninfected rats avoid cat odors, but rats infected with *T. gondii* are drawn to this scent, which may increase transmission to feline hosts.

Roles in Ecosystems

Modifying the behavior of infected hosts, to make transmission to other hosts more likely to occur, is one way parasites can affect the structure of ecosystems. For example, in the case of *Euhaplorchis californiensis* (discussed above) it is plausible that the local predator and prey species might be different if this parasite were absent from the system.

Although parasites are often omitted in depictions of food webs, they usually occupy the top position. Parasites can function like keystone species, reducing the dominance of superior competitors and allowing competing species to co-exist.

Many parasites require multiple hosts of the different species to complete their life cycles and rely on predator-prey or other stable ecological interactions to get from one host to another. In this sense, the parasites in an ecosystem reflect the health of that system.

Value

Although parasites are generally considered to be harmful, the eradication of all parasites would not necessarily be beneficial. Parasites account for as much as or more than half of life's diversity; they perform an important ecological role (by weakening prey) that ecosystems would take some time to adapt to; and without parasites, organisms may eventually tend to asexual reproduction, diminishing the diversity of sexually dimorphic traits. Parasites provide an opportunity for the transfer of genetic material between species. On rare, but significant, occasions this may facilitate evolutionary changes that would not otherwise occur, or that would otherwise take even longer.

References

- Morris PJ, Salt C, Raila J, Brenten T, Kohn B, Schweigert FJ, Zentek J. Safety evaluation of vitamin A in growing dogs. British Journal of Nutrition. 2012; 108(10):1800-1809

- D. E. C. Corbridge (1995). Phosphorus: An Outline of its Chemistry, Biochemistry, and Technology (5th ed.). Amsterdam: Elsevier. ISBN 0-444-89307-5

- Moran, N.A. (2006), "Symbiosis", Current Biology, 16 (20): 866–871, PMID 17055966, doi:10.1016/j.cub.2006.09.019, retrieved 2007-09-23

- Goldwater, William. "Analysis of Adipose Tissue in relation to Body Weight Loss in Man". Journal of Applied Physiology. Retrieved June 28, 2011

- Ullrey, D. E. (2004). "Nutrient Requirements: Carnivores". In Pond, Wilson. Encyclopedia of Animal Science. CRC Press. p. 670. ISBN 978-0-8247-5496-9

- Labandeira, C.C. (2005). "The four phases of plant-arthropod associations in deep time" (PDF). Geologica Acta. 4 (4): 409–438. Archived from the original (Free full text) on 26 June 2008. Retrieved 15 May 2008

- Labandeira, C.C. (1998). "Early History Of Arthropod And Vascular Plant Associations 1". Annual Review of Earth and Planetary Sciences. 26 (1): 329–377. doi:10.1146/annurev.earth.26.1.329

- Nugent, G; Challies, CN (1988). "Diet and food preferences of white-tailed deer in north-eastern Stewart Island". New Zealand Journal of Ecology. 11: 61–73

- Thomas, Peter & Packham, John. Ecology of Woodlands and Forests: Description, Dynamics and Diversity. Publisher: Cambridge University Press 2007. ISBN 978-0521834520

- "Tree Squirrels". The Humane Society of the United States. Archived from the original on December 25, 2008. Retrieved 2009-01-01

- Getz, W (2011). "Biomass transformation webs provide a unified approach to consumer–resource modelling". Ecology Letters. 14: 113–124. PMC 3032891. PMID 21199247. doi:10.1111/j.1461-0248.2010.01566.x

- Langenheim, J.H. (1994). "Higher plant terpenoids: a phytocentric overview of their ecological roles". Journal of Chemical Ecology. 20 (6): 1223–1280. doi:10.1007/BF02059809

- Herrera, C.M. (1985). "Determinants of Plant-Animal Coevolution: The Case of Mutualistic Dispersal of Seeds by Vertebrates". Oikos. 44 (1): 132–141. doi:10.2307/3544054

- Hutson, Jarod M.; Burke, Chrissina C.; Haynes, Gary (2013-12-01). "Osteophagia and bone modifications by giraffe and other large ungulates". Journal of Archaeological Science. 40 (12): 4139–4149. doi:10.1016/j.jas.2013.06.004

- Collocott, T. C. (ed.) (1974). Chambers Dictionary of science and technology. Edinburgh: W. and R. Chambers. ISBN 0-550-13202-3. CS1 maint: Extra text: authors list (link)

- Annex: Towards a Forestry Commission England Grey Squirrel Policy (PDF), UK: Forestry Commission, 22 Jan 2006, retrieved 15 May 2012

- Karban, R.; Agrawal, A.A. (2002). "Herbivore Offense". Annual Review of Ecology and Systematics. 33: 641–664. doi:10.1146/annurev.ecolsys.33.010802.150443

- Sterner, Robert W.; Elser, James J.; and Vitousek, Peter. Ecological Stoichiometry: The Biology of Elements from Molecules to the Biosphere. Publisher: Princeton University Press 2002. ISBN 978-0691074917

- Foley, James A. (February 7, 2014). "Carnivorous, Pre-Dinosaur Predator was First to Evolve Steak Knife-like Teeth". Nature World News. Retrieved May 3, 2014

- Ullrey, D. E. (2004). "Mammals: Carnivores". In Pond, Wilson. Encyclopedia of Animal Science. CRC Press. p. 591. ISBN 978-0-8247-5496-9

Permissions

We would like to thank the editorial team for lending their expertise to make the book truly unique. They have played a crucial role in the development of this book. Without their invaluable contributions this book wouldn't have been possible. They have made vital efforts to compile up to date information on the varied aspects of this subject to make this book a valuable addition to the collection of many professionals and students.

This book was conceptualized with the vision of imparting up-to-date and integrated information in this field. To ensure the same, a matchless editorial board was set up. Every individual on the board went through rigorous rounds of assessment to prove their worth. After which they invested a large part of their time researching and compiling the most relevant data for our readers.

The editorial board has been involved in producing this book since its inception. They have spent rigorous hours researching and exploring the diverse topics which have resulted in the successful publishing of this book. They have passed on their knowledge of decades through this book. To expedite this challenging task, the publisher supported the team at every step. A small team of assistant editors was also appointed to further simplify the editing procedure and attain best results for the readers.

Apart from the editorial board, the designing team has also invested a significant amount of their time in understanding the subject and creating the most relevant covers. They scrutinized every image to scout for the most suitable representation of the subject and create an appropriate cover for the book.

The publishing team has been an ardent support to the editorial, designing and production team. Their endless efforts to recruit the best for this project, has resulted in the accomplishment of this book. They are a veteran in the field of academics and their pool of knowledge is as vast as their experience in printing. Their expertise and guidance has proved useful at every step. Their uncompromising quality standards have made this book an exceptional effort. Their encouragement from time to time has been an inspiration for everyone.

The publisher and the editorial board hope that this book will prove to be a valuable piece of knowledge for students, practitioners and scholars across the globe.

Index

www.ingramcontent.com/pod-product-compliance
Lightning Source LLC
Chambersburg PA
CBHW082039190326
41458CB00010B/3414